D0082748

DISCARDED
WIDENER UNIVERSITY

Results and Problems in
Combinatorial Geometry

RESULTS AND PROBLEMS IN COMBINATORIAL GEOMETRY

V. G. Boltjansky
and
I. Ts. Gohberg

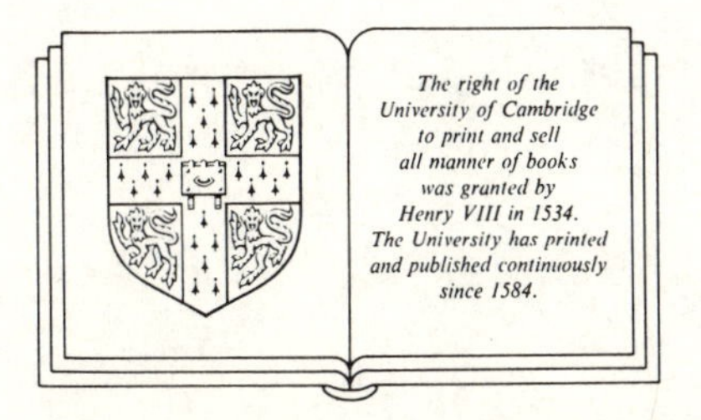

CAMBRIDGE UNIVERSITY PRESS
Cambridge
London New York New Rochelle
Melbourne Sydney

Published by The Press Syndicate of the University of Cambridge
The Pitt Building, Trumpington Street, Cambridge CB2 1RP
32 East 57th Street, New York, NY 10022, USA
10 Stamford Road, Oakleigh, Melbourne 3166, Australia

First published 1985

Printed in Great Britain at the University Press, Cambridge

Library of Congress catalogue card number: 85-4187

British Library cataloguing in publication data

Boltjansky, V. G.

Results and problems in combinatorial geometry.
1. Combinatorial geometry
I. Title II. Gohberg, Izrail III. Teoremy i zadachi kombinatornoi geometrii. *English*
516'.13 QA167

ISBN 0 521 26298 4
ISBN 0 521 26923 7 Pbk

CONTENTS

FOREWORD

There are many elegant results in the theory of convex bodies that may be fully understood by high school students, and at the same time be of interest to expert mathematicians. The aim of this book is to present some of these results. We shall discuss combinatorial problems of the theory of convex bodies, mainly connected with the partition of a set into smaller parts.

The theorems and problems in the book are fairly recent; the oldest of them is just over thirty years old, and many of the theorems are still in their infancy. They were published in professional mathematical journals during the last five years.

We consider the main part of the book to be suitable for high school students interested in mathematics. The material indicated as complicated may be skipped by them. The most straightforward sections concern plane sets: §§1-3, 7-10, 12-14. The remaining sections relate to spatial (and even *n*-dimensional) sets. For the keen and well-prepared reader, at the end of the book will be found notes, as well as a list of journals, papers and books. References to the notes are given in round brackets, ( ), and references to the bibliography in square brackets, [ ]. In several places, especially in the notes, the discussion is at the level of scientific papers. We did not consider it inappropriate to include such material in a non-specialized book. We feel that it is possible to popularize science, not only for the layman but also for the benefit of the expert.

The book brings the reader up-to-date as far as the problems considered here are concerned. At the end of the book (§19) some unsolved problems are stated. Several of them are so intuitive and so easy to state that even able high school students can speculate about their solutions.

In conclusion, let us say a few words about combinatorial geometry itself. This is a new branch of geometry which is not yet in its final form; it is too early to speak of combinatorial geometry as a subject apart. Apart from the problems presented in this book, a group of problems connected with Helly's theorem (see Chapter 2 [37]) are without doubt related to combinatorial geometry, as are problems about packings and coverings of sets (see the excellent book by Fejes Tóth [12]), as well as a series of other problems. For the interested reader, we also very much recommend the book by Hadwiger and Debrunner [24], devoted to problems of combinatorial geometry of the plane, and the most interesting paper of Grünbaum [18], closely connected with the material presented to the reader.

The authors would like to take this opportunity to express their sincere gratitude to I.M. Yaglom, whose enthusiasm and friendly participation greatly contributed to improving the text of this book.

V.G. Boltjansky

I.Ts. Gohberg

INTRODUCTION TO THE ENGLISH TRANSLATION

This book originally appeared in Russian almost twenty years ago; nevertheless it is as fresh now as then. No better exposition of the main results has since appeared, and the problems stated at the end of the book still remain unsolved.

I would like to mention two books which appeared after this volume and which are closely related to this material. The first is "The Decomposition of Figures into Smaller Parts" by the same authors, which appeared in English translation in the University of Chicago Press in 1980, and also the book of V. G. Boltyansky and P. S. Soltan "Combinatorial Geometry of Different Classes of Convex Sets" Stiintsa, Kishinev, 1978 (in Russian). The first book is a popular book devoted only to combinatorial problems of the plane, and the second book is on the level of mathematical research monographs.

Finally, I would like to thank Cambridge University Press and Dr. David Tranah for their interest and cooperation.

I. Gohberg

Tel Aviv
20th November, 1984

CHAPTER 1

PARTITION OF A SET INTO SETS OF SMALLER DIAMETER

§1. THE DIAMETER OF A SET

Consider a disc of diameter d. Any two points M and N of this disc (fig. 1) are at distance at most d, and the disc also contains two points A and B whose distance is exactly d.

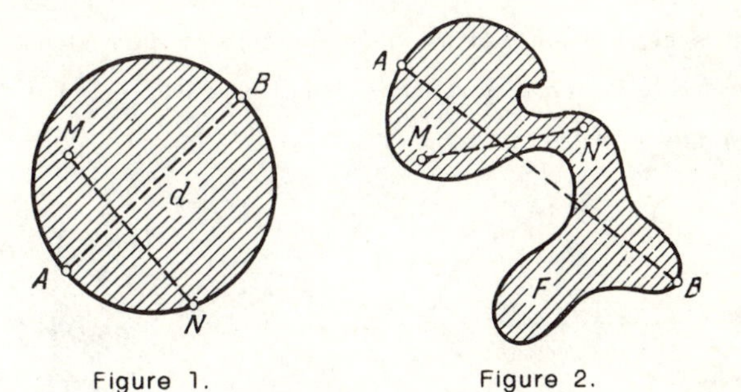

Figure 1. Figure 2.

Now consider another set instead of the disc. What can one call the "diameter" of this set? The observation above leads to the definition of the diameter of a set as the *greatest* distance between its points. In other words, we say that a *set* F (fig. 2) *has diameter* d *if, firstly, any two points* M *and* N *of* F *are at distance at most* d*, and secondly, one can find at least two points* A *and* B *whose distance is exactly* d (1).

For example, let F be a half-disc (fig. 3). Denote by A and B the endpoints of the semicircular arc. Then it is clear that the diameter of F is the length of the segment AB. In general, if F is a circular segment bounded by an arc ℓ and a chord a, then if the arc ℓ is not greater than a semicircle (fig. 4*a*), the diameter of F equals a (that is, the length of a chord), and if ℓ is greater than a semicircle (fig. 4*b*), then the diameter of F is the same as the diameter of the entire disc.

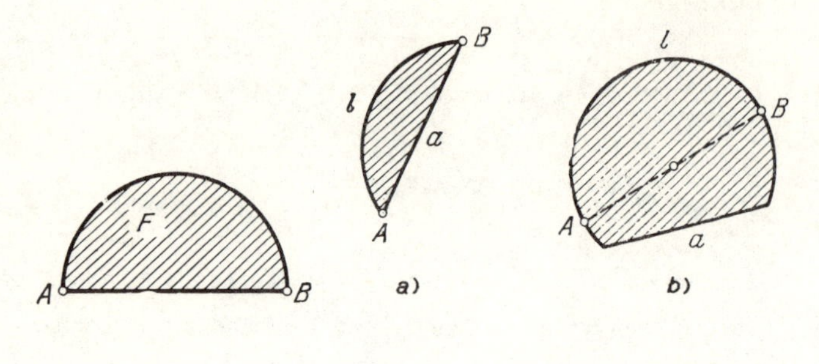

Figure 3. Figure 4.

It is easily seen that the diameter of a polygon F (fig. 5) is the maximal distance among its vertices. In particular, the diameter of a triangle is the length of a longest side (fig. 6).

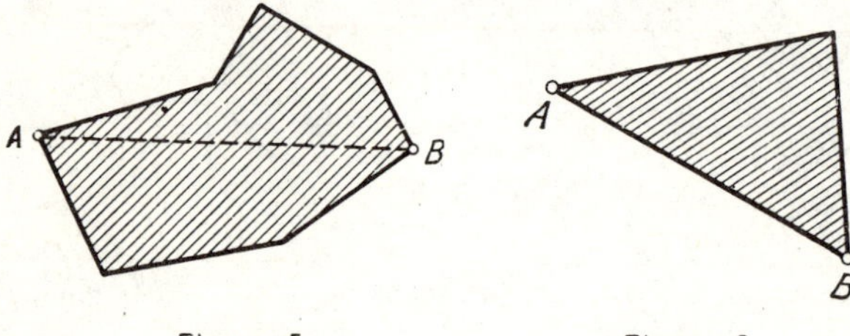

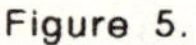
Figure 5.

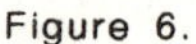
Figure 6.

Note that a set F of diameter d may contain many pairs of points at distance d. For example, an ellipse (fig. 7) contains only one such pair, a square (fig. 8) contains two pairs, an equilateral triangle (fig. 9) contains three pairs and, lastly, a disc contains infinitely many such pairs.

§2. THE PROBLEM

It is easily seen that if a disc of diameter d is partitioned into two parts by some curve MN, then at least one of these parts has diameter d. Indeed, if M' is the point diametrically opposite M, then it must belong to one of the parts, and this part (containing M

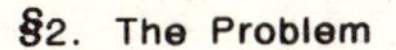

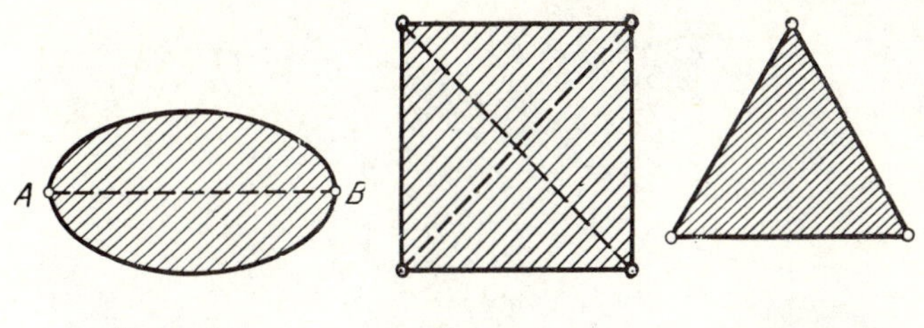

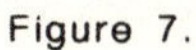

Figure 7. Figure 8. Figure 9.

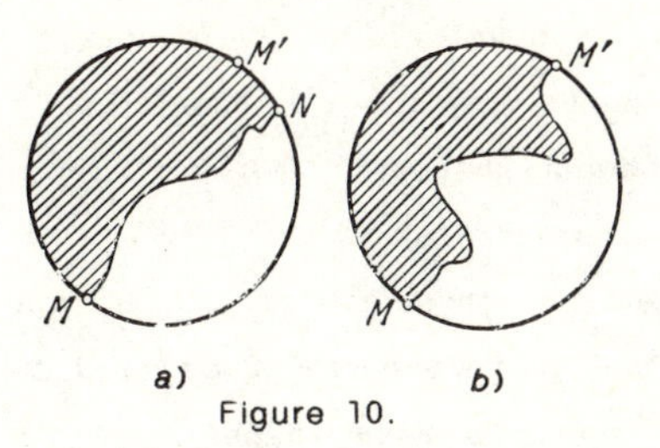

a) b)

Figure 10.

and M') has diameter d (fig. 10) (2). Furthermore, it is clear that the disc can be partitioned into three parts each of diameter less than d (fig. 11).

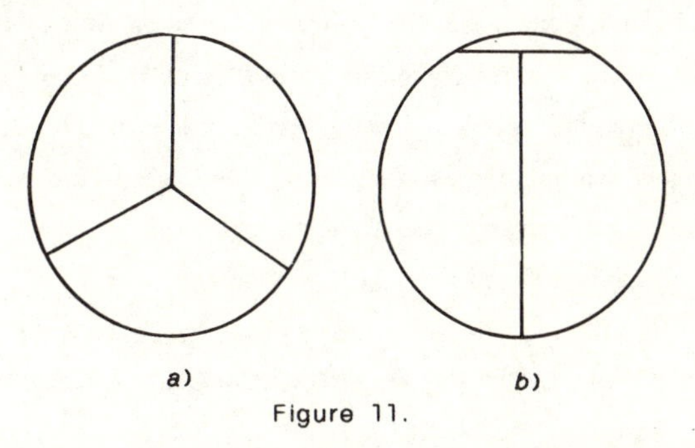

a) b)

Figure 11.

Thus, a disc of diameter d cannot be partitioned into two parts of diameter less than d, but can be partitioned into three such parts. The same holds for an equilateral triangle of side d (for if it is partitioned into two parts, one of the parts will contain at least two vertices of the triangle, and this part will have diameter d). However, there are sets that can be partitioned into two sets of smaller diameter (fig. 12).

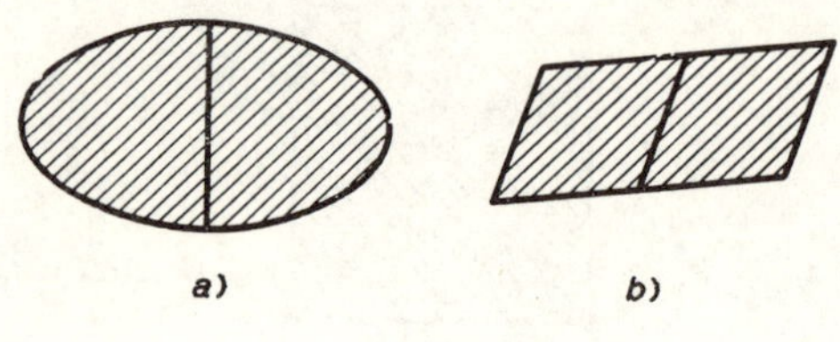

Figure 12.

Given a set F we can consider the problem of partitioning it into parts of smaller diameter (3). We denote by $a(F)$ the minimal number of sets needed in such a partition. Thus, if F is a disc or an equilateral triangle, then $a(F) = 3$, and for an ellipse or for a parallelogram we have $a(F) = 2$.

The problem of partitioning a set into sets of smaller diameter can be generalised from plane sets to bodies in three-dimensional space (or even in n-dimensional space, if the reader is familiar with this concept).

The problem of finding the possible values of $a(F)$ was posed in 1933 by the well-known Polish mathematician K. Borsuk [4]. Since then, numerous research papers have dealt with this problem. The results obtained are presented in the first chapter of this book.

Firstly we shall consider plane sets, then present a solution for three-dimensional bodies and, finally, we review the results in the n-dimensional case for the well-prepared reader.

§3. A SOLUTION OF THE PROBLEM FOR PLANE SETS

We have seen that $a(F)$ is 2 for some plane sets, and is 3 for some others. The question arises whether one can find a plane set F with $a(F) > 3$, that is, a set for which there is no partition into three parts of smaller diameter, and one has to use four or more parts. It turns out that three parts indeed always suffice, that is, we have the following theorem, proved by Borsuk in 1933 [4].

<u>Theorem 1</u>. *Given a plane set F of diameter d, $a(F) \leqslant 3$; that is, F can be partitioned into three parts of diameter less than d.*

Proof. The main part of the proof is the following lemma, proved in 1920 by the Hungarian mathematician J. Pál [33]: *every plane set of diameter d can be surrounded by a regular hexagon whose opposite sides are at distance d* (fig. 13).

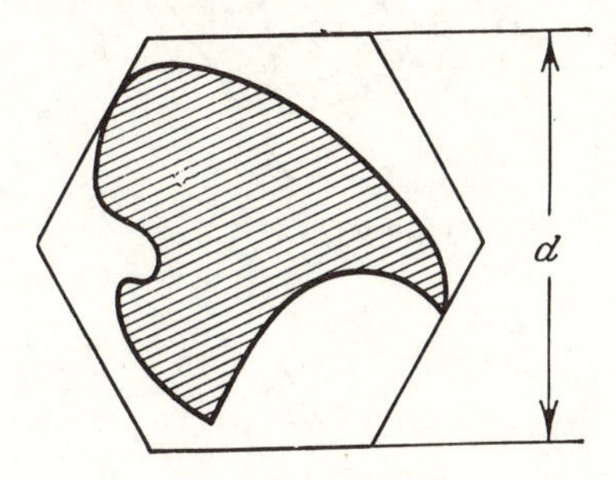

Figure 13.

Take a line ℓ that does not intersect the set F, and move it closer to F (keeping it parallel to its original direction), until it touches F (fig. 14). The resulting line ℓ_1 has at least one point in common with F, and the whole set F lies on one side of ℓ_1. Such a line is called a *support line* of F (4). Let us draw a second support line ℓ_2, parallel to ℓ_1 (fig. 14). Clearly, the whole set F will lie in the strip between the lines ℓ_1 and ℓ_2, and the distance between the lines is at most d (since the diameter of F is d).

Now draw two support lines m_1, m_2 at an angle of 60° to ℓ_1 (fig. 15). The lines ℓ_1, ℓ_2, m_1, m_2 form a parallelogram $ABCD$ with angle 60° and heights at most d, surrounding the set F.

Next draw two support lines p_1, p_2 at an angle of 120° to ℓ_1, and denote by M and N the bases of the perpendiculars dropped on these lines from the ends of the diagonal AC (fig. 15). We shall show that the direction of ℓ_1 can be chosen so that that $AM = CN$. Indeed, suppose $AM \neq CN$, say $AM < CN$. Then the value $y = AM-CN$ is negative. Now, we rotate ℓ_1 through 180° (the

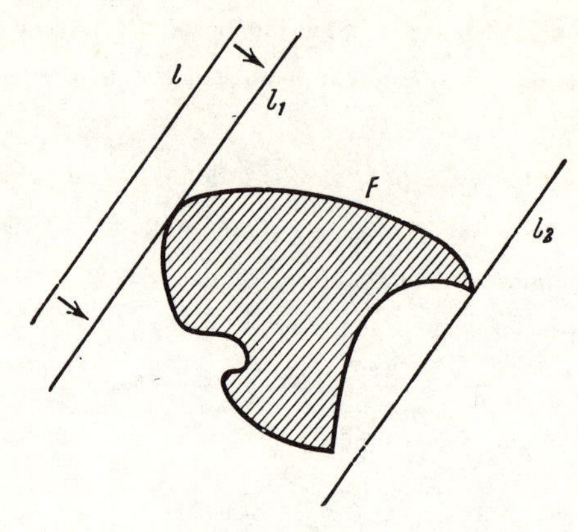

Figure 14.

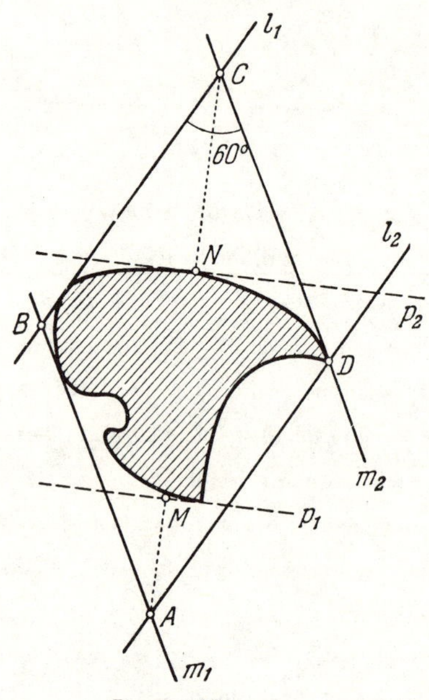

Figure 15.

set F is kept fixed). The remaining lines ℓ_2, m_1, m_2, p_1, p_2 will also change their positions (since their positions are determined by the choice of ℓ_1). Therefore, as ℓ_1 rotates, the points A, C, M, N (5) will continuously move and continuously vary the value of

$y = AM - CN$. But when the line ℓ_1 has rotated through 180°, it will lie in the position formerly occupied by ℓ_2. Hence, we shall obtain the same parallelogram as in Figure 15 with the points A and C, and also M and N, reversed. Consequently, y will be positive. If we now plot the graph of the rotation of ℓ_1 from 0° to 180° (fig. 16), we see that y is zero for some position of ℓ_1, i.e. $AM = CN$

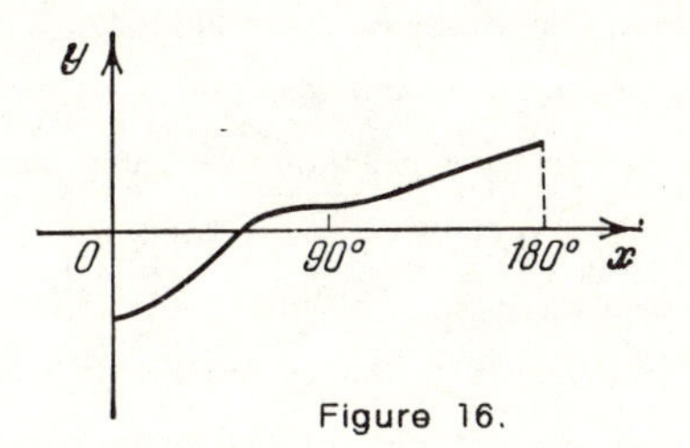

Figure 16.

(since as y continuously changes from negative to positive, it must at some point be zero). We shall examine the positions of all our lines when y is zero (fig. 17). The equality $AM = CN$ implies that

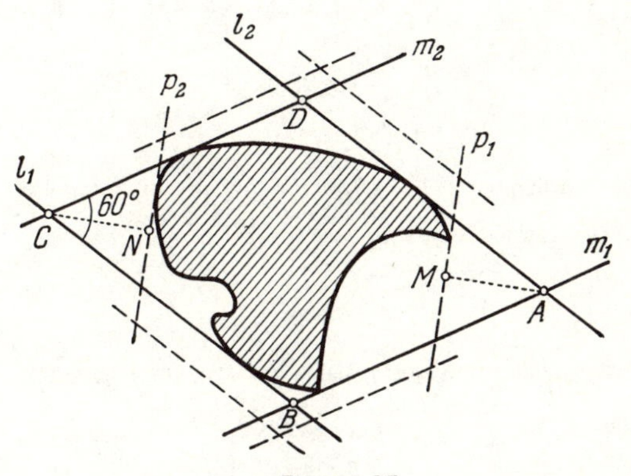

Figure 17.

the hexagon formed by the lines ℓ_1, ℓ_2, m_1, m_2, p_1, p_2 is centrally symmetric. Each angle of this hexagon is 120°, and the distance between opposite sides is at most d. If the distance between the lines p_1 and p_2 is less than d, we shall move them apart (moving each the same distance) until the distance equals d.

We then move the lines ℓ_1, ℓ_2 and m_1, m_2 in exactly the same way. We thereby obtain a centrally symmetric hexagon (with angles 120°), with opposite sides at distance d from each other (the dotted hexagon in fig. 17). From the above, it is clear that all the sides of this hexagon are equal, that is, the hexagon is regular with the set F lying inside.

Now we show that it is possible to partition this regular hexagon into three parts, each having diameter less than d. In addition, the set F will also be partitioned into three parts, each of diameter less than d. The required partition of the regular hexagon into three parts is shown in Figure 18 (the points P, Q and R are the centres of the sides, and O is the centre of the hexagon). The diameters of the parts are less than d since in the triangle PQL, the angle Q is a right-angle, and so $PQ < PL = d$.

Figure 18.

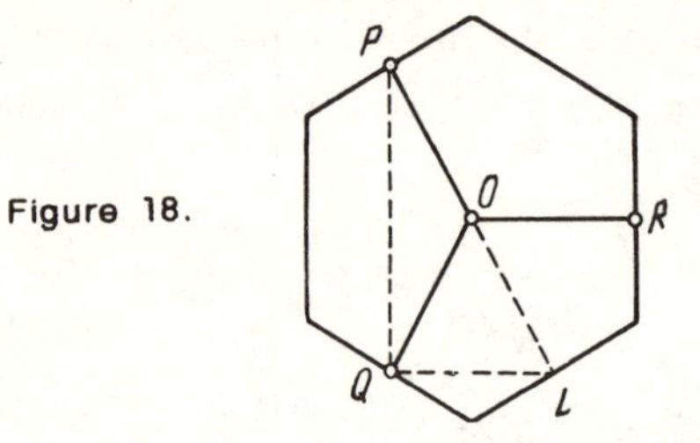

This proves Theorem 1. (See Problem 4 in connection with this.)

§4. PARTITION OF A BALL INTO PARTS OF SMALLER DIAMETER

It is easily seen that in three-dimensional space there exist bodies F for which $a(F)$ equals 2 or 3. For example, if the body is very elongated in one direction (fig. 19*a*), then $a(F) = 2$ (fig. 19*b*). Furthermore, if F is a cone with height less than the radius of the base (fig. 20*a*), then $a(F) = 3$. In fact, the diameter of this body equals the diameter of the base, and therefore, $a(F) \geqslant 3$

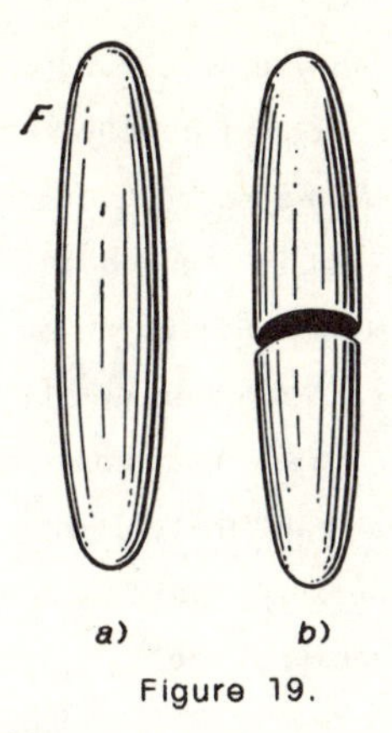

Figure 19.

(because it is impossible to partition even the disc at the base into two parts of smaller diameter); the partition of F into three parts of smaller diameter is shown in Figure 20*b*.

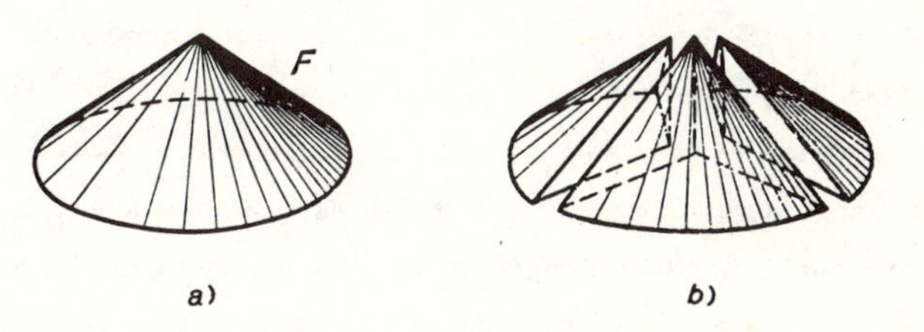

Figure 20.

It turns out that in space, there exist bodies for which $a(F) > 3$. For example, a regular tetrahedron with side d has this property (if it is partitioned into three parts, one of the parts must contain two vertices of the tetrahedron, and therefore, the diameter of this part is d). Theorem 2 which follows shows the significantly deeper fact that a ball is also such a body.

<u>Theorem 2</u>. *A ball of diameter d cannot be partitioned into three parts, each of which has diameter less than d.*

Before moving to the proof, let us compare this theorem with what has already been said. (The reader not familiar with the concept of "n-dimensional" may proceed to the proof of Theorem 2 or even skip the proof and proceed directly to section §5 or Chapter 2.) As we have seen, it is impossible to partition the disc into two parts of smaller diameter. Let us call the disc a *two-dimensional ball* (two-dimensional because it lies in the plane which, as is well-known, has two dimensions). We then get the following assertion: *it is impossible to partition a two-dimensional ball into two parts of smaller diameter.* The usual ball (that is, lying in three-space) is naturally called a *three-dimensional ball*. Combining the cases of the disc and the ball, we get the following:

Theorem 2′. *For $n = 2$ or 3, it is impossible to partition an n-dimensional ball into n parts of smaller diameter.*

Apart from two-space (that is, the plane) and three-space, in mathematics and its applications, spaces of four and more dimensions are also considered. It turns out that Theorem 2′ holds not only for $n = 2$ or 3, but for an arbitrary natural number n. This theorem in its general form was proved by K. Borsuk [3] in 1932, but the essence of this result, though stated differently, was obtained even earlier (in 1930) by the Soviet mathematicians L.A. Lyusternik and L.G. Shnirel'man [32]. The proofs found by these mathematicians are highly complicated and sophisticated (they are based on theorems related to a branch of geometry called *topology*), and hence cannot be presented here. However, for $n = 3$, there is an elementary proof. (See also the theorems mentioned on page 83, proved by the German mathematician H. Lenz.)

Proof of Theorem 2. Let E be a ball of diameter d. Suppose, contrary to the assertion, that it is possible to partition E into three parts M_1, M_2, M_3, each of which has diameter less than d. Let S be the surface of the ball E. Denote by N_1 the set of all points of S belonging to M_1, and define N_2 and N_3 analogously. The sphere S is thus partitioned into three parts N_1, N_2, N_3, each of which clearly has diameter less than d. Let d_1 be the diameter of N_1 (so $d_1 < d$), and put $h = (d-d_1)/3$.

Now perform the following construction on the sphere S. Choose two diametrically opposite points P and Q (the poles of S), and intersect S by several planes perpendicular to the line PQ. These planes intersect S in parallel circles, dividing S into "polar caps" and several belts. We shall divide each of these belts into several parts by arcs of meridians, thereby getting a partition of the surface resembling brickwork (fig. 21a). Furthermore, let us choose the number of meridians and parallels to be large enough to

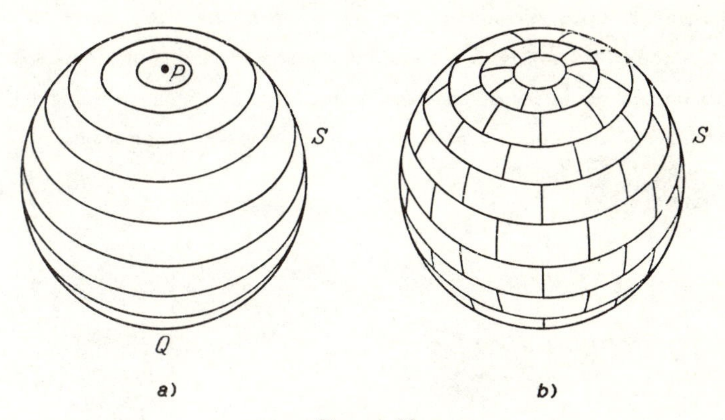

a) b)

Figure 21.

ensure that each of the parts into which the surface is divided (the polar caps and the bricks) has diameter less than h.

Consider now each of the parts having a common point with the set N_1. Taken together, they form a set which we shall denote

Figure 22.

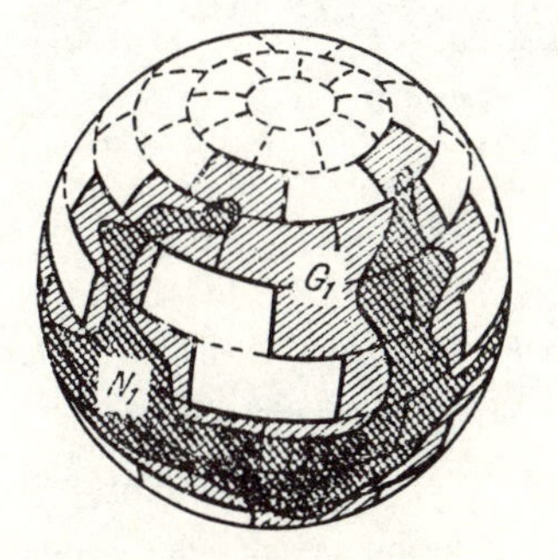

by G_1. As N_1 has diameter d_1 and the diameter of each of the parts is less than h, the diameter of G_1 is less than $d_1 + 2h$. But:

$$d_1 + 2h = d - h < d$$

so the diameter of G_1 is less than d.

Now consider the boundary of G_1. It is easy to see that it consists of a finite number of closed curves, which intersect neither themselves nor each other (fig. 22). In fact, at each point where there is a junction, only three parts meet (fig. 21*b*). If the point of a junction lies on the boundary of G_1, then of the three adjoining parts, one (fig. 23) or two (fig. 24) belong to G_1. Take now any point on the boundary of the set G_1 and begin to move it along the boundary. The boundary of G goes along a well-defined path until

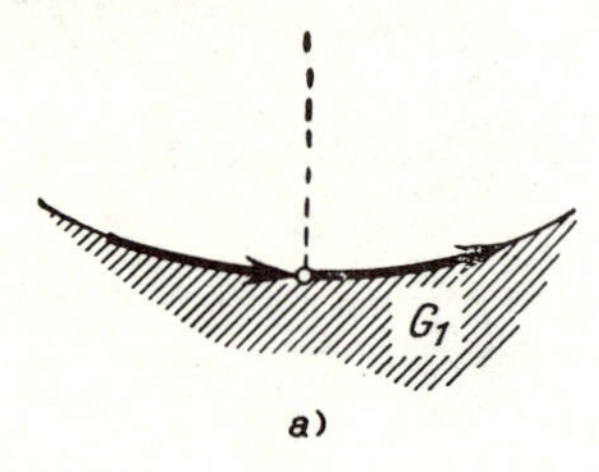

a)

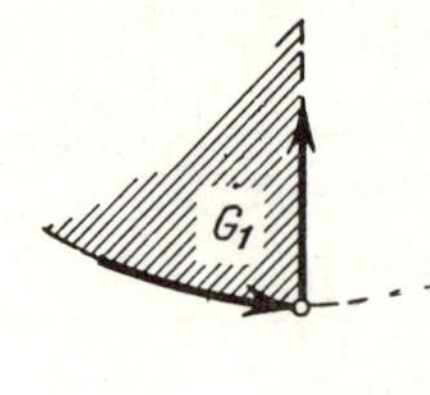

b)

Figure 23.

a)

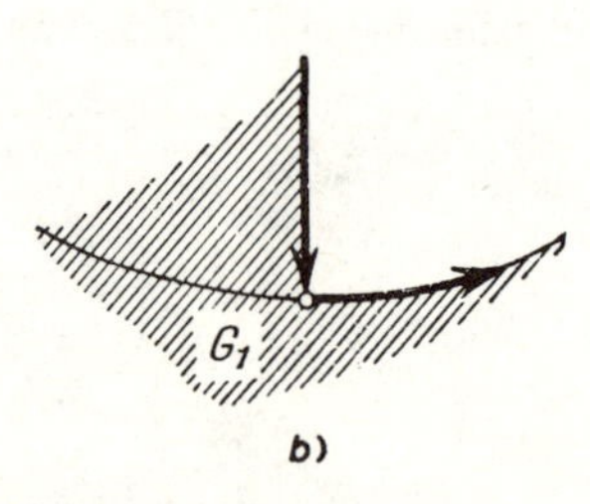

b)

Figure 24.

we reach a junction. But even there, the boundary does not split, but proceeds further in the same manner (this is immediate from Figures 23 and 24). As there is only a finite number of parts, then by going further and further along the boundary of G_1, we must inevitably return to the starting point, that is, describe a closed curve (as the boundary line cannot terminate anywhere). Notice, however, that the boundary of G_1 may consist not only of one straight line, but of several (fig. 22). We shall denote the closed lines forming the boundary of G_1 by $L_1, L_2, \ldots, L_k$.

Now let G_1' be the set symmetric to G_1 with respect to the centre of S, that is, G_1' consists of all the points of S diametrically opposite the points of G_1. It is easily seen that the sets G_1 and G_1' do not have any points in common. In fact, if the point A were to belong to both G_1 and G_1', then the point B diametrically opposite A would belong to G_1 (since A belongs to G_1'). But then G_1 would contain two diametrically opposite points A and B, contradicting the fact that G_1 has diameter less than d.

The boundary G_1' is formed by the lines $L_1', L_2', \ldots, L_k'$, symmetric to the lines $L_1, L_2, \ldots, L_k$. As the sets G_1 and G_2 do not have any common points, the closed lines $L_1, L_2, \ldots, L_k, L_1', L_2', \ldots, L_k'$ do not intersect each other pairwise.

Now notice that if, on the sphere S, we are given q closed lines which intersect neither themselves nor each other, then they divide the surface into $q+1$ parts. This is easy to see by induction: one line divides the surface into two parts, and each subsequent added line forms one new part (6).

As we have $2k$ lines $L_1, L_2, \ldots, L_k, L_1', L_2', \ldots, L_k'$, they divide the surface into $2k+1$ parts, that is, an odd number of parts. We shall call these parts *countries*. Each country is either wholly contained in G_1 or in G_1' or lies outside both G_1 and G_1'. As the lines $L_1, L_2, \ldots, L_k$ are symmetric to the lines $L_1', L_2', \ldots, L_k'$, each country either has its symmetric country, or is self-symmetric with respect to the centre of the sphere. The number of countries pairwise symmetric to each other is even, and as the total number of countries is odd, at least one country can be found which is self-symmetric with respect to the centre of the sphere. Let H be such a country and C be one of its interior points. As the country H is self-symmetric, the point C' lying diametrically opposite C also belongs to H. From this it is clear that the diameter of H is d, and therefore all the interior of H lies outside G_1 and G_1'. But as H is one country, it is represented by a whole, connected part of the sphere, and therefore, the points C and C' (fig. 25) can be joined by a path Γ wholly lying inside H. The path Γ', symmetric to Γ, joins the same points C and C', and also lies wholly within H. Γ and Γ' have no common points with the set G_1, and moreover, have no common points with N_1.

Figure 25.

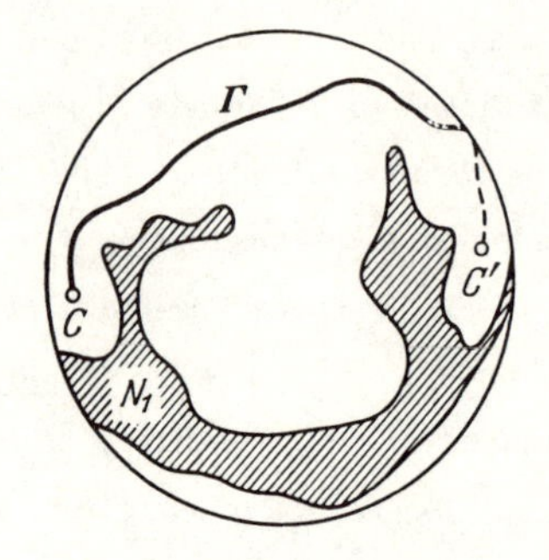

Let us now return to the sets N_1, N_2, N_3 mentioned at the beginning of the proof. Each point of the path Γ belongs to at least one of the sets N_2, N_3. The endpoints C and C' (as they are diametrically opposite) belong to different sets N_2 and N_3; without loss of generality, let C belong to N_2 and C' to N_3. We shall move along Γ from C to C' and denote by D the last point meeting the set N_2 (fig. 26). If D does not belong to N_3, then neither do

Figure 26.

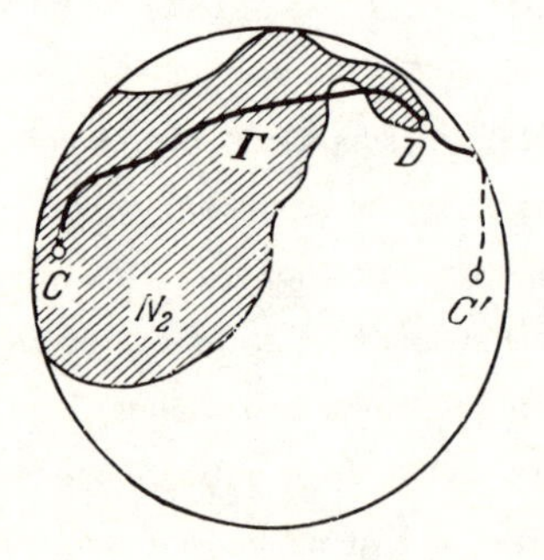

the points near to D (7). But then the points lying on Γ between D and C' and close to D cannot belong to any of the sets N_1, N_2, N_3, which is impossible. Hence, the point D belongs to both N_2 and N_3.

Lastly, consider the point D' diametrically opposite D. It belongs to the path Γ', and consequently is not contained in N_1. But neither N_2 nor N_3 contain it, since these sets have diameter less than d and contain the point D. Thus, the point D' is not contained in any of the sets N_1, N_2, N_3, contradicting the hypothesis.

This contradiction shows that it is impossible to partition the ball E into three parts of smaller diameter, completing the proof of Theorem 2.

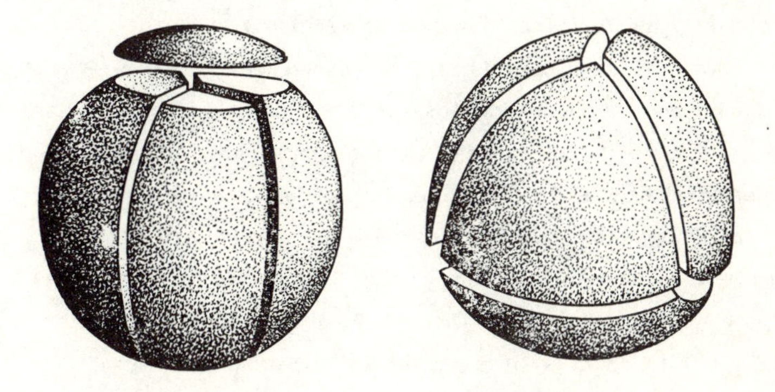

Figure 27. Figure 28.

According to the result above for a ball E we have $a(E) > 3$. What, in fact, is the value of $a(E)$? Can a ball be partitioned into four parts of smaller diameter, or is a larger number of parts required? It is easy to see that $a(E) = 4$, that is, a ball can be partitioned into four parts of smaller diameter. One such partition is shown in Figure 27. Another, more symmetric, partition may be obtained as follows. Inscribe in the ball E of diameter d a regular tetrahedron $ABCD$, and consider the solid angles $OABC$, $OABD$,

OACD and *OBCD* with common vertex *O*, where *O* is the centre of the tetrahedron. These four solid angles cut the ball *E* into four parts (fig. 28), each of which has diameter less than *d*.

§5. A SOLUTION FOR THREE-DIMENSIONAL BODIES

This section is concerned with proving the following theorem:

Theorem 3. *Let F be a three-dimensional body of diameter d. Then* $a(F) \leqslant 4$, *that is, F can be partitioned into four parts of smaller diameter.*

Before proceeding to the proof, let us make a few remarks about the place of this theorem in combinatorial geometry, and about the history of its appearance and proof. (These *n*-dimensional arguments may also be skipped.)

We have seen that, for any two-dimensional set *F*, $a(F) \leqslant 3$, and moreover, for a two-dimensional ball (that is, a disc), this inequality becomes an equality. At the same time, for the three-dimensional ball, $a(E) = 4$. Thus, if we denote the *n*-dimensional ball by E^n (where $n = 2, 3$), we have the equality $a(E^n) = n + 1$. This relation holds not only for $n = 2, 3$, but also for an arbitrary natural number n. In fact, Theorem 2′ above states that $a(E^n) \geqslant n + 1$, that is, it is impossible to partition the ball E^n into *n* parts of smaller diameter. At the same time, $n+1$ parts are sufficient; this is established by the construction in *n*-dimensional space of a partition of the ball E^n analogous to the partition for $n = 3$ in Figure 27. We shall not go into this in detail here. For the reader familiar with *n*-dimensional geometry, the construction of the partitions analogous to those in Figures 27 and 28 will not be particularly difficult.

So, $a(E^n) = n + 1$. But for $n = 2$, the two-dimensional ball E^2 (that is, the disc), is one of the sets which requires the maximum number of parts for a partition into parts of smaller diameter, that is, it is one of the sets for which the inequality $a(F) \leqslant 3$ attains equality. It is natural to conjecture that this situation remains the case for all larger values of n. This conjecture was stated by K. Borsuk [4] in 1933. In other words, Borsuk conjectured the following:

Borsuk's conjecture. *For any n-dimensional body F of diameter d,* $a(F) \leqslant n+1$*; that is, F may be partitioned into* $n+1$ *parts of smaller diameter.*

The efforts of many mathematicians around the world were directed towards proving this conjecture. However, it took a long time to find a complete solution even for $n = 3$, that is, for bodies in normal three-space. Such a solution was obtained in 1955 by the English mathematician H.G. Eggleston [7]. He showed that Borsuk's conjecture is true in three-dimensional space, that is, Theorem 3 holds.

It should be noted that the original proof due to Eggleston was very complicated, long and difficult. In 1957, the Israeli mathematician B. Grünbaum proposed a new, shorter, and very elegant proof of this Theorem [15]. The ideas resemble those used in the proof of Theorem 1: a body F is surrounded by a certain polytope which is then partitioned into four parts of diameter less than d. In what follows, we shall present Grünbaum's proof.

Proof of Theorem 3. The first part of the proof will follow from the following lemma, proved in 1953 by the American mathematician D. Gale [13]: *every three-dimensional body F of diameter d may be surrounded by a right octahedron whose opposite faces are at distance d.*

Consider the right tetrahedron $ABCA'B'C'$ which has A and A', B and B', C and C' as pairwise opposite vertices, the distance between the opposite faces being d (fig. 29). All the eight faces of the octahedron are pairwise parallel. We shall not consider all four pairs of parallel planes in which these faces lie, but only three of them; for example, take the planes ABC' and $A'B'C$, $AB'C$ and

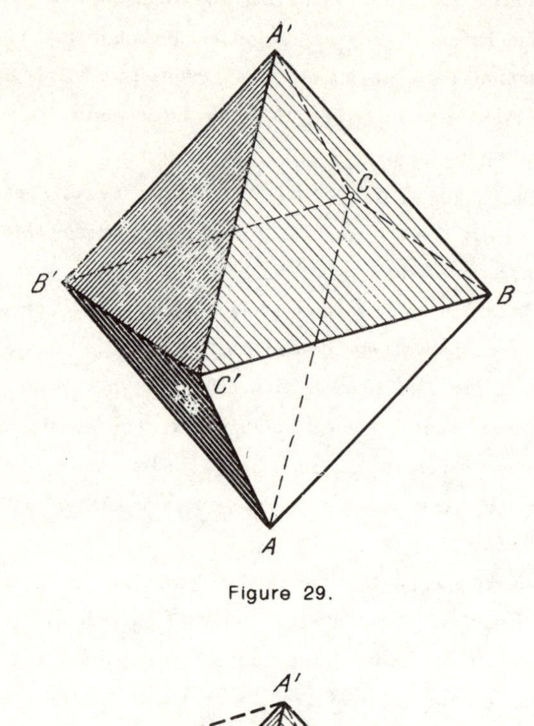

Figure 29.

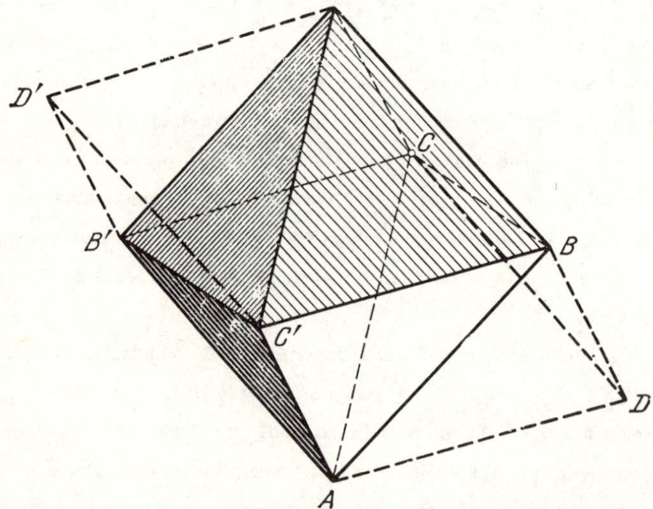

Figure 30.

$A'BC'$, $A'BC$ and $AB'C'$. These three pairs of parallel planes intersecting each other form the parallelepiped $AB'CDA'BC'D$ (see Figure 30, in which the new edges of the parallelepiped are shown by heavy dotted lines); we shall denote this parallelepiped by Φ. The distance between the opposite faces of the parallelepiped is, as before, equal to d. Furthermore, the diagonal DD' is perpendicular to the discarded faces ABC and $A'B'C'$ of the octahedron. Thus, the parallelepiped Φ has the property that if two planes are perpendicular to the diagonal DD' and are at a distance $d/2$ from the centre of the parallelepiped, then they cut off two triangular pyramids, and the remaining middle part is a right octahedron. Let us also observe that the plane $BDB'D'$ is a plane of symmetry of the parallelepiped Φ, and the line ℓ, perpendicular to this plane and passing through the centre of the diagonal DD', is its axis of symmetry. In other words, if the parallelepiped is rotated about ℓ by 180°, it will be in the same position (fig. 31).

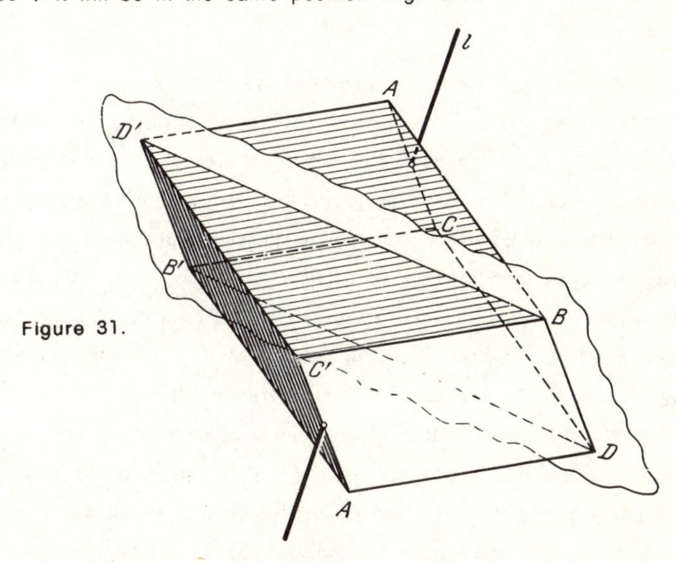

Figure 31.

Now let F be a body of diameter d. Draw two planes, parallel to the face $AB'CD$ of the parallelepiped Φ, so that the body F lies between them (fig. 32). Then begin to bring these planes towards the body F, keeping them all the time parallel to $AB'CD$, until they

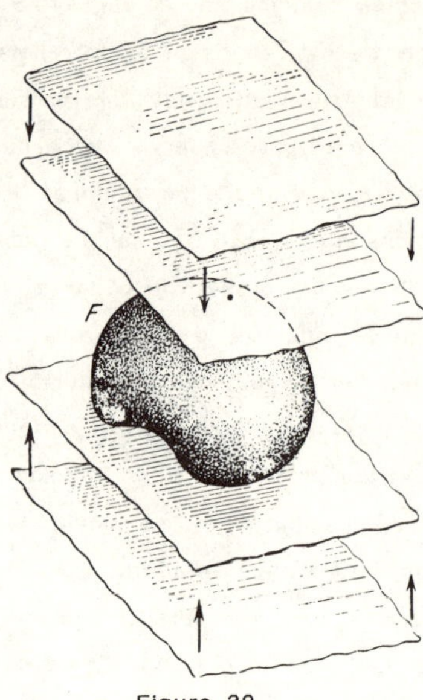

Figure 32.

touch F. We thus get two support planes of the body, parallel to $AB'CD$. Then construct two more pairs of support planes parallel to the other faces of the parallelepiped. As a result, a parallelepiped is constructed which encloses F and has faces parallel to the faces of Φ. We shall denote this enclosing parallelepiped by Π, and its diagonal corresponding to DD' by EE'. Draw two more support planes of F, perpendicular to the diagonal DD' of Φ. Denote the perpendiculars dropped from the points E and E' onto these planes by EM and $E'M'$, and let y be the difference $EM-E'M'$.

We shall show that it is possible to position the initial parallelepiped Φ in space so that $EM = E'M'$. In fact, let us assume that $EM \neq E'M'$; without loss of generality, let $EM < E'M'$, so $y = EM-E'M'$ is negative. Now continuously rotate Φ around ℓ through 180°, (when, consequently, it occupies the same position as before). The parallelepiped Π will also continuously change with Φ, as will the support planes perpendicular to the diagonal DD'.

Therefore, the points E, E' and M, M' will be continuously displaced as Φ rotates, and consequently will continuously change the value of $y = EM - E'M'$. After a rotation through 180°, the points E and E' will have changed places, and so y will be positive. Portraying graphically the dependence of y on the angle of rotation as in Figure 11, we see that there exists an angle of rotation of Φ at which y vanishes, that is, $EM = E'M'$. We shall consider this position of the parallelepiped Φ (and Π). Let α and β denote the support planes perpendicular to the diagonal DD'.

If the distance between any two opposite faces of Π is less than d, move the planes of these faces apart (withdrawing them the same distance from the centre of the parallelepiped), so that the distance between them equals d. We similarly deal with all three pairs of parallel faces of Π, and also the parallel planes α and β. As a result, we obtain a new parallelepiped Π^*, equal to the initial parallelepiped Φ, and two planes α^* and β^* perpendicular to the diagonal DD', lying at distance $d/2$ from the centre of Π^*. These planes cut off two triangular pyramids from Π^*, and the remaining part is represented by a right octahedron. It is clear that the body F lies inside this octahedron.

So we have surrounded the body F having diameter d by the right octahedron $ABCA'B'C'$, which has opposite faces at distance d apart.

The next part of the proof will be concerned with the construction of a polytope V, somewhat smaller than the polytope $ABCA'B'C'$, and also containing the body F. Thus, draw two planes γ and γ' perpendicular to the diagonal AA' and lying at distance $d/2$ from the centre of the octahedron. These two planes cut off two pyramids (with apexes A and A') from the octahedron. It is easy to see that the interior of one of the pyramids does not contain any points of F (because if P and Q are interior points of these pyramids, they are situated on opposite sides of the region bounded by the planes γ and γ', and so $PQ > d$). We may suppose without loss of generality that the interior of the pyramid with apex A' does not contain points of F (otherwise A and A' could be swapped).

The polytope remaining from the octahedron after the removal of the pyramid with apex A' wholly contains the body F (fig. 33).

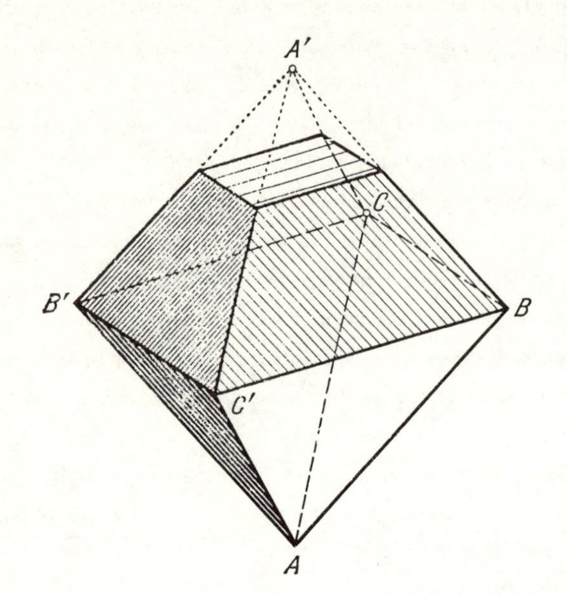

Figure 33.

Now we construct two planes perpendicular to the diagonal BB' and situated at distance $d/2$ from the centre of the octahedron. They again cut off two pyramids (with apexes B and B') and, moreover, the interior of one of these pyramids does not contain points of F. Without loss of generality, let this be the pyramid with apex B' (fig. 34). The polytope obtained from the previous one after the deletion of the pyramid with apex B' also contains the body F. Analogously, it is possible to cut off one of the similar pyramids with apexes C and C'; let this be, without loss of generality, the pyramid with apex C'. We arrive at the polytope V, shown in Figure 35, which also contains F.

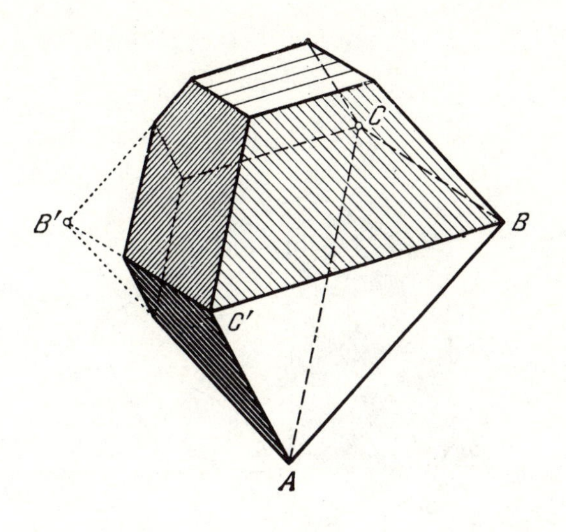

Figure 34.

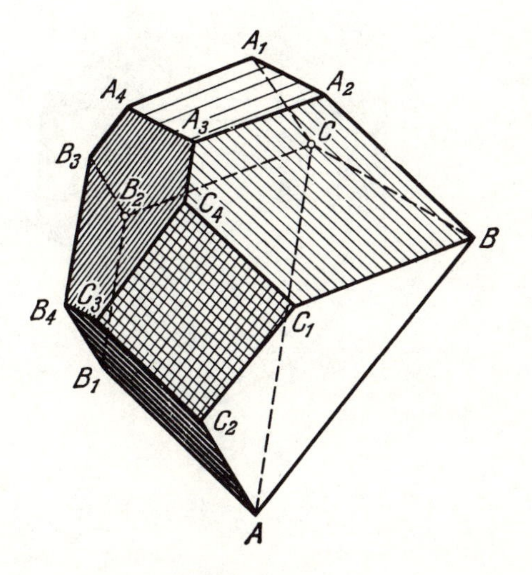

Figure 35.

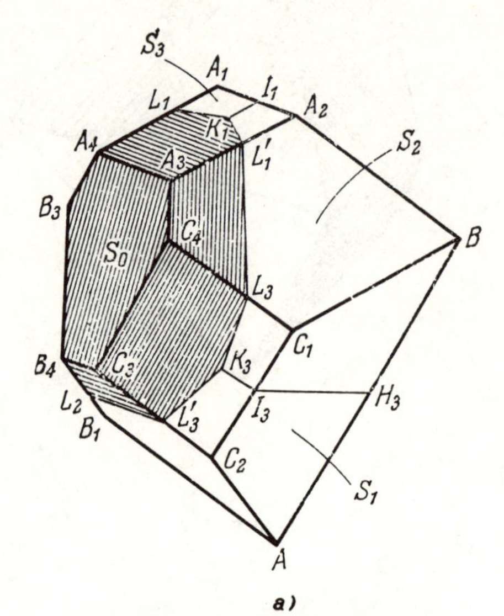

S_3
A_1
I_1
A_2
L_1
K_1
S_2
A_4
A_3
L_1'
B_3
C_4
B
S_0
L_3
C_1
B_4
C_3
K_3
L_2
I_3
H_3
B_1
L_3'
C_2
S_1
A

a)

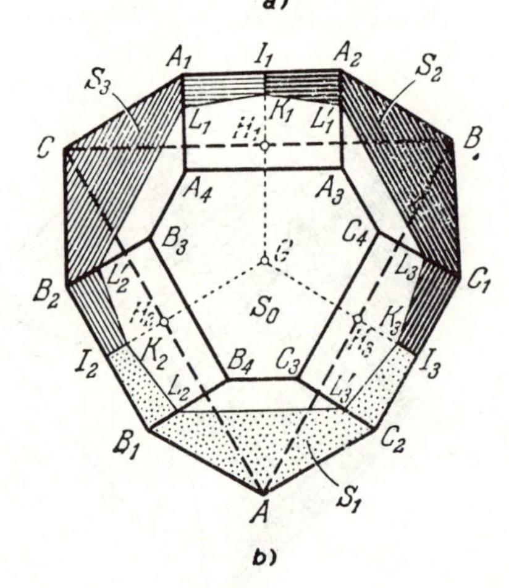

A_1
I_1
A_2
S_3
S_2
L_1
H_1
K_1
L_1'
C
B
A_4
A_3
B_3
C_4
G
L_3
C_1
B_2
L_2'
S_0
H_2
K_3
H_3
K_2
I_2
B_4
C_3
I_3
L_2
L_3'
B_1
C_2
S_1
A

b)

Figure 36.

may be partitioned into four parts, each of which has diameter less than d (because F, which is surrounded by V, will then be partitioned into four parts, the diameter of each of which is, a fortiori, less than d). Let us construct such a subdivision of V (see fig. 36*a* and fig. 36*b* showing a picture of the polytope V from the side of a hexagonal face). V has one triangular face ABC (remaining from the octahedron), three square faces $A_1A_2A_3A_4$, $B_1B_2B_3B_4$, $C_1C_2C_3C_4$ (the bases of the cut pyramids), three pentagonal faces, and three trapezoidal faces. Let G be the centre of the equilateral triangle ABC, H_1, H_2, H_3 be the centres of the sides of this triangle, and I_1, I_2, I_3 be the centres of the small bases of the trapezia. Take some points K_1, K_2, K_3 lying in the quadrilateral faces, and some points L_1, L'_1, L_2, L'_2, L_3, L'_3 lying on the lateral sides of the squares (not parallel to the bases of the trapezia). Joining the chosen points, we partition the surface of the polytope V into four regions S_0, S_1, S_2, S_3, bounded by the closed broken lines

$$L'_1K_1L_1L'_2K_2L_2L'_3K_3L_3L'_1, \qquad GH_2I_2K_2L_2L'_3K_3I_3H_3G,$$
$$GH_3I_3K_3L_3L'_1K_1I_1H_1G, \qquad GH_1I_1K_1L_1L'_2K_2I_2H_2G.$$

Denote now by O the centre of the octahedron obtained from the polytope V by cutting off pyramids. Consider all segments joining O to the points of the region S_0. All such segments fill some body V_0, represented as a "pyramid" with apex O, and "base" the non-planar region S_0. We analogously construct the bodies V_1, V_2, V_3 as "pyramids" with apex O and "bases" S_1, S_2, S_3. Together, the bodies V_0, V_1, V_2, V_3 make up the whole polytope V (fig. 37).

Up to now, we have not fixed the exact positions of the points K_1, K_2, K_3 and L_1, L'_1, L_2, L'_2, L_3, L_3 on the square faces and their sides. We shall now choose these points in such a way that each of the bodies V_0, V_1, V_2, V_3 has diameter less than d. Namely, we shall choose the points L_1, L'_1, L_2, L'_2, L_3, L'_3 so that

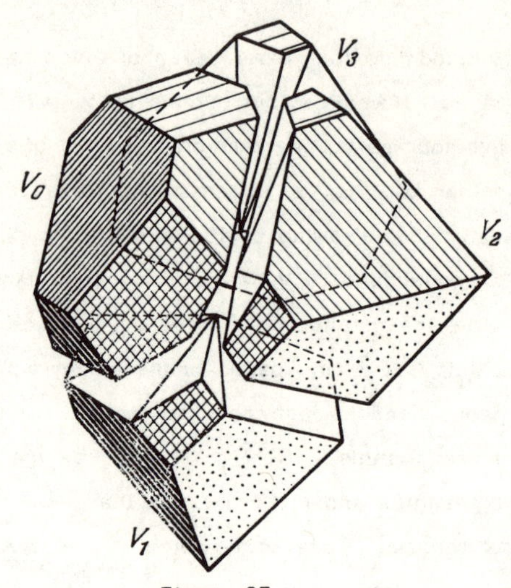

Figure 37.

they are at a distance of

$$\frac{15\sqrt{3}-10}{46\sqrt{2}}\, d$$

from the smaller bases of the trapezia (that is, so that the indicated path has segments A_1L_1, A_2L_1', ...). Furthermore, choose the point K_1 so that $K_1L_1 = K_1L_1'$, and so that the distance from the point K_1 to the smaller base of the trapezium (that is, to the segment A_1A_2) equals

$$\frac{1231\sqrt{3}-1986}{1518\sqrt{2}}\, d.$$

The reader should not be surprised at the complexity of the choice of these numbers. They have been found with the help of complicated calculations in Grünbaum's proof (these numbers were chosen so that all the parts V_0, V_1, V_2, V_3 have the same diameter). It turns out that for such a choice of points K_1, K_2, K_3, L_1, L_1', L_2, L_2', L_3, L_3', the diameter of each of the bodies V_0, V_1, V_2, V_3 is in fact less than unity, namely, each has diameter:

$$\frac{\sqrt{6129030-937419\sqrt{3}}}{1518\sqrt{2}}\, d \approx 0.9887d.$$

To prove this result, let us just say that to evaluate the diameter of the polytope V_0, it is necessary to find all possible distances between its vertices and choose the largest of them. Solving this problem is elementary, (for example, with the help of multiple applications of the theorem), but it involves tedious computation. By means of this computation (printed below; we recommend that it be skipped at a first reading), we complete the proof of Theorem 3.

Let us take a rectangular system of coordinates Oxyz and six points:

$A\ (a,0,0)$ $\qquad B\ (0,a,0)$ $\qquad C\ (0,0,a)$

$A'\ (-a,0,0)$ $\qquad B'\ (0,-a,0)$ $\qquad C'\ (0,0,-a)$

where a is positive. These six points are the vertices of a right octahedron. It is clear that the plane in which the face ABC of this octahedron lies has equation $x+y+z = a$; this plane lies at a distance $a/\sqrt{3}$ from the centre of the octahedron (that is, from the origin of the coordinates). Consequently, the distance d between two parallel faces of this octahedron is given by $d = 2a/\sqrt{3}$.

The plane, perpendicular to the diagonal AA', is parallel to the plane Oyz. Thus, the plane perpendicular to the diagonal AA' situated at distance $d/2$ from the centre of the octahedron and cutting off the pyramid with apex A', has equation $x = -d/2$. From here, it is easy to find the coordinates of the points A_1, A_2, A_3, A_4 (fig. 35). For example, A_2 lies in the plane Oxy (that is, the plane $z = 0$), in the plane $x = -d/2$, and in the plane of the face $A'BC$, that is, in the plane with equation:

$$-x + y + z = a = d\sqrt{3}/2$$

Consequently, the point A_2 has coordinates:

$$x = -a/\sqrt{3} = -d/2, \qquad y = a - (a/\sqrt{3}) = d(\sqrt{3}-1)/2, \qquad z = 0$$

Analogously, let us find the coordinates of all the points A_i, B_i, C_i:

A_1 $(-d/2, 0, b)$ A_2 $(-d/2, b, 0)$

A_3 $(-d/2, 0, -b)$ A_4 $(-d/2, -b, 0)$

B_1 $(b, -d/2, 0)$ B_2 $(0, -d/2, b)$

B_3 $(-b, -d/2, 0)$ B_4 $(0, -d/2, -b)$

C_1 $(0, b, -d/2)$ C_2 $(b, 0, -d/2)$

C_3 $(0, -b, -d/2)$ C_4 $(-b, 0, -d/2)$

where b denotes $a - a/\sqrt{3} = d(\sqrt{3} - 1)/2$. Thus, the coordinates of all vertices of the polytope V are computed.

Let us proceed to calculate the coordinates of the vertices of the polytopes V_0, V_1, V_2, V_3. The point G has coordinates:

$$x = y = z = \frac{a}{3} = \frac{d}{2\sqrt{3}};$$

$$G \quad \left[\frac{d}{2\sqrt{3}}, \frac{d}{2\sqrt{3}}, \frac{d}{2\sqrt{3}}\right]$$

The points H_1, H_2, H_3 are easily found as the centres of the segments BC, CA, AB:

$$H_1 \quad \left[0, \frac{d\sqrt{3}}{4}, \frac{d\sqrt{3}}{4}\right], \qquad H_2 \quad \left[\frac{d\sqrt{3}}{4}, 0, \frac{d\sqrt{3}}{4}\right],$$

$$H_3 \quad \left[\frac{d\sqrt{3}}{4}, \frac{d\sqrt{3}}{4}, 0\right]$$

Furthermore, the points I_1, I_2, I_3 are the centres of the segments A_1A_2, B_1B_2, C_1C_2:

$$I_1 \quad \left[-\frac{d}{2}, \frac{b}{2}, \frac{b}{2}\right], \qquad I_2 \quad \left[\frac{b}{2}, -\frac{d}{2}, \frac{b}{2}\right],$$

$$I_3 \quad \left[\frac{b}{2}, \frac{b}{2}, -\frac{d}{2}\right]$$

Let us now determine the coordinates of the points L_1 and L_1'. The vector p, directed from A_1 to A_4 and having length 1 has the form:

$$p = \left\{0, \frac{-1}{\sqrt{2}}, \frac{-1}{\sqrt{2}}\right\}.$$

Therefore, $\overline{A_1L_1} = \overline{A_2L'_1} = cp$, where:

$$c = \frac{15\sqrt{3}-10}{46\sqrt{2}}\, d.$$

This enables us to determine easily the coordinates of the points L_1, L'_1:

$$L_1 \left[-\frac{d}{2}, \frac{-c}{\sqrt{2}}, b - \frac{c}{\sqrt{2}} \right], \qquad L'_1 \left[-\frac{d}{2}, b - \frac{c}{\sqrt{2}}, \frac{-c}{\sqrt{2}} \right]$$

We analogously find the remaining points L_i and L'_i:

$$L_2 \left[b - \frac{c}{\sqrt{2}}, -\frac{d}{2}, \frac{-c}{\sqrt{2}} \right], \qquad L'_2 \left[\frac{-c}{\sqrt{2}}, -\frac{d}{2}, b - \frac{c}{\sqrt{2}} \right],$$

$$L_3 \left[\frac{-c}{\sqrt{2}}, b - \frac{c}{\sqrt{2}}, -\frac{d}{2} \right], \qquad L'_3 \left[b - \frac{c}{\sqrt{2}}, \frac{-c}{\sqrt{2}}, -\frac{d}{2} \right]$$

Lastly, by the definition of the point K_1 we have $\overline{I_1K_1} = ep$, where

$$e = \frac{1231\sqrt{3}-1986}{1518\sqrt{2}}\, d.$$

From here, we find the coordinates of the point K_1 (and analogously the points K_2, K_3):

$$K_1 \left[-\frac{d}{2}, \frac{b}{2} - \frac{e}{\sqrt{2}}, \frac{b}{2} - \frac{e}{\sqrt{2}} \right], \qquad K_2 \left[\frac{b}{2} - \frac{e}{\sqrt{2}}, -\frac{d}{2}, \frac{b}{2} - \frac{e}{\sqrt{2}} \right]$$

$$K_3 \left[\frac{b}{2} - \frac{e}{\sqrt{2}}, \frac{b}{2} - \frac{e}{\sqrt{2}}, -\frac{d}{2} \right]$$

By the same token, all the vertices of the polytopes V_0, V_1, V_2, V_3 are determined (the one common vertex of these polytopes lies at the origin).

Now, in order to determine the diameter of the polytope V_0 (or V_1, V_2, V_3), it is necessary to find the maximum of the distances between its vertices. This is easily done as the coordinates of all the vertices are known. For example, knowing the points

$$K_1 = \left[-\frac{d}{2}, \frac{b}{2} - \frac{e}{\sqrt{2}}, \frac{b}{2} - \frac{e}{\sqrt{2}} \right]$$

$$= \left[-\frac{d}{2}, \frac{1227-472\sqrt{3}}{1518}\cdot\frac{d}{2}, \frac{1227-472\sqrt{3}}{1518}\cdot\frac{d}{2} \right],$$

$$L_2 = \left[b - \frac{c}{\sqrt{2}}, -\frac{d}{2}, \frac{-c}{\sqrt{2}}\right]$$

$$= \left[\frac{31\sqrt{3}-36}{46}\cdot\frac{d}{2}, -\frac{d}{2}, -\frac{15\sqrt{3}-10}{46}\cdot\frac{d}{2}\right],$$

we easily find that the length of the segment K_1L_2 equals:

$$\sqrt{\left[b - \frac{c}{\sqrt{2}} + \frac{d}{2}\right]^2 + \left[-\frac{d}{2} - \frac{b}{2} + \frac{e}{\sqrt{2}}\right]^2 + \left[-\frac{c}{\sqrt{2}} - \frac{b}{2} + \frac{e}{\sqrt{2}}\right]^2}$$

or, substituting in the values of b, c and e we get

$$\frac{\sqrt{6129030-937419\sqrt{3}}}{1518\sqrt{2}}\, d \approx 0.9887d.$$

This is the maximum of the distances between the vertices of the polytope V_0 (that is, the diameter of V_0; see page 27). The diameters of the polytopes V_1, V_2, V_3 are calculated similarly.

We notice that in this proof, the polytope V is partitioned into parts V_0, V_1, V_2, V_3, the diameters of which differ very slightly from d. Naturally this occurs because the polytope V contains not only the body F, but also much "spare space".

If the polytope V had been selected more economically, it would have been possible to decrease somewhat the bound $0.9887d$ estimating the sizes of the parts (see Problem 4 in connection with this).

We point out that the solution of Borsuk's problem in three-dimensional space was given by the Hungarian mathematician A. Heppes [25] simultaneously with Grünbaum. However, his proof is less well-known, as it is published in Hungarian which is not known by most mathematicians.* In Heppes' solution, the partition into parts is less economical than in the proof given. He obtained a bound of $0.9977d$ for the diameter of the parts.

---

*Note added in Translation: This paper exists in German also.

§6. BORSUK'S CONJECTURE FOR *N*-DIMENSIONAL BODIES*

The reader is now obviously interested in what the situation is concerning the proof of Borsuk's conjecture in spaces of more than three dimensions. Unfortunately, this problem in its general form is still not solved, in spite of the efforts of many mathematicians. It is not even known whether it is true for bodies lying in four-dimensional space, that is, it is not known whether any four-dimensional body of diameter d may be partitioned into five parts of smaller diameter. In this is contained one of the interesting features of the problem we are considering; the sharp contrast between the extreme simplicity of the statement of the problem, and the huge difficulties in its solution, which seem at present to be completely insurmountable. (See Problems 1, 2, 3, 5 in connection with this.)

However, for some special kinds of n-dimensional body, the validity of Borsuk's conjecture has been established.

In the first place, we mention the work of the well-known Swiss geometer H. Hadwiger. Hadwiger does not consider arbitrary n-dimensional bodies, but only convex ones (the reader will find a few words about convex sets in Section 7), because it is clearly sufficient to prove Borsuk's conjecture for convex bodies (see page 43). In one of his papers in 1946, Hadwiger considered n-dimensional convex bodies with smooth boundary, that is, convex bodies which have a natural support hyperplane across each boundary point. By an elegant argument, Hadwiger showed that for such convex bodies, Borsuk's conjecture is true. In other words, we have the following:

Theorem 4. *Every n-dimensional convex body with smooth boundary and diameter d may be partitioned into $n + 1$ parts of diameter less than d.*

---

*We recommend that the reader not familiar with n-dimensional geometry move straight to Chapter 2.

Proof. Let F be any n-dimensional convex body with smooth boundary having diameter d. Consider also an n-dimensional ball E having the same diameter d, and construct some partition of this ball E into $n+1$ parts of diameter less than d (see Figures 27 and 28). We shall denote the parts into which E is partitioned by $M_0, M_1, \ldots, M_n$. We now construct a partition of the boundary G of the body F into $n+1$ sets $N_0, N_1, \ldots, N_n$. Let A be an arbitrary boundary point of F.

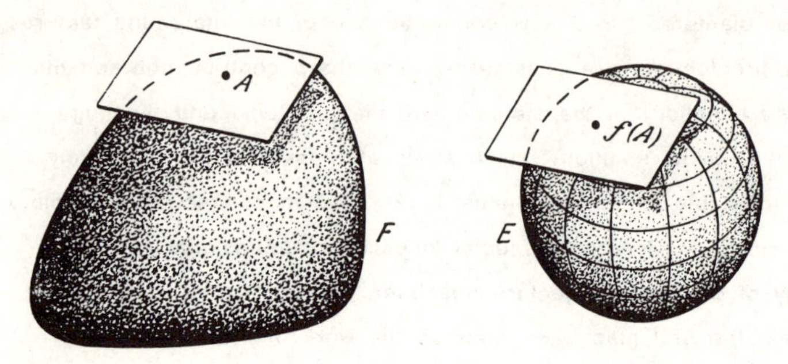

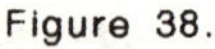

Figure 38.

Draw the support hyperplane of F passing through A (this is, by hypothesis, unique), and draw parallel to it the tangential hyperplane of the ball E, so that the body F and the ball E lie on the same side of these hyperplanes (fig. 38). Denote by $f(A)$ the point at which the constructed hyperplane touches the ball E. We shall consider the point A belonging to the set N_i if the corresponding point $f(A)$ belongs to the set M_i ($i = 0, 1, \ldots, n$). Consequently, the whole boundary G of the body F is partitioned into $n+1$ sets $N_0, N_1, \ldots, N_n$ (8).

We shall prove that each of the sets $N_0, N_1, \ldots, N_n$ has diameter less than d. Let us assume that contrary to this, a certain set N_i has diameter d, and let A and B be two points of the set N_i at distance d from each other. Construct two hyperplanes Γ_A, Γ_B passing through the points A and B and perpendicular to the segment AB. Clearly, F lies in the region between these hyperplanes (otherwise the diameter of F would be greater than d), that is, Γ_A and Γ_B are support hyperplanes of F, passing through A and B. These support planes being parallel implies that the corresponding points $f(A)$ and $f(B)$ lying on the boundary of the ball E are diametrically opposite, that is, the distance between the points $f(A)$ and $f(B)$ equals d. On the other hand, as A and B belong to the set N_i, the points A and B also belong to M_i, and therefore the distance between $f(A)$ and $f(B)$ is less than d. This contradiction shows that none of the sets $N_0, N_1, \ldots, N_n$ has diameter d.

Now let O be an arbitrary interior point of F. For any $i = 0, 1, \ldots, n$, we shall denote by P_i the "cone" with apex O and curvilinear base the set N_i. Clearly the constructed "cones" $P_0, P_1, \ldots, P_n$ fill the whole body F, that is, we have obtained a partition of F into $n+1$ parts. Furthermore, it is clear that each of the sets P_i has diameter less than d (because the diameter of the "base" N_i is less than d). Hence, the constructed partition divides the body F into $n+1$ parts of diameter less than d, proving Theorem 4.

In another paper in 1947, refining the above proof, Hadwiger proved the following theorem:

If an n-dimensional convex body of diameter d is such that a small n-dimensional ball of radius r may freely roll inside the convex boundary of this body, then this convex body can be partitioned into n+1 parts, the diameter of each of which does not exceed:

$$d - 2r\left[1 - \sqrt{(1 - 1/n^2)}\right]$$

So for convex bodies with a smooth boundary, Borsuk's conjecture is true (Theorem 4). There remain convex bodies having corners (that is, points at which the support plane is not unique). For such bodies, there are up to now practically no results. However in 1955 the German mathematician H. Lenz showed that *any n-dimensional convex body may be partitioned into parts of smaller*

diameter, the number of which does not exceed $(\sqrt{n}+1)^n$.*
However, this bound is, of course, not exact, and is rather far from Borsuk's conjecture. For example, Lenz's bound guarantees the possibility of partitioning any four-dimensional body into 81 parts of smaller diameter, while Borsuk's conjecture requires that a partition of a four-dimensional body into five parts of smaller diameter be shown possible! The latest result is by L. Danzer [5], who gave a stronger bound:

$$a(F) < \sqrt{\frac{(n+2)^3}{3}} \cdot (2+\sqrt{2})^{(n-1)/2}$$

(For a four-dimensional body, this bound establishes the possibility of a partition into 55 parts of smaller diameter!)

**Proof.* Let us denote by m the integer satisfying the inequality:

$$\sqrt{n} < m \leqslant \sqrt{n} + 1$$

Furthermore, let us enclose the n-dimensional body F of diameter d in a cube with side d, and partition each of the edges of this cube into m equal parts. Drawing through these points a division of the hyperplane, parallel to the faces of the cube, we partition the cube into m^n smaller cubes with side d/m. The diameter of each of these cubes equals $d\sqrt{n}/m$ and is therefore less than d:

$$\frac{d}{m}\sqrt{n} < \frac{d}{\sqrt{n}}\sqrt{n} = d$$

The constructed partition into small cubes induces a partition of the body F into parts of diameter less than d, and moreover, the number of these parts does not exceed m^n, that is, does not exceed $(\sqrt{n}+1)^n$.

CHAPTER 2

COVERING CONVEX BODIES WITH HOMOTHETIC BODIES AND THE ILLUMINATION PROBLEM

§7. CONVEX SETS

A plane set F is called *convex* if, whenever it contains two points, it contains the whole segment joining them (fig. 39). Thus,

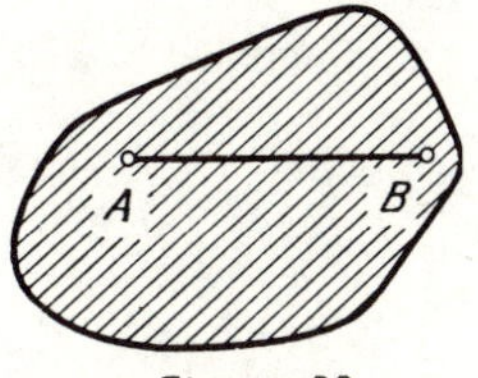

Figure 39.

for example, the triangle, parallelogram, trapezium, disc, segment of a disc and the ellipse are examples of convex sets (fig. 40). In Figure 41 are examples of non-convex sets. The sets shown in Figure 40 are bounded. There exist also unbounded (extending to infinity) convex sets: a half-plane, an angle less than 180° etc. (fig. 42).

The points of any convex set F partition into two classes, *interior* points and *boundary* points. Points which are surrounded on all sides by points of F are regarded as interior points. Thus, if A is an interior point of F, then a disc of some radius (even if very small) with centre at A belongs wholly to F (fig. 43). At a boundary point of F, there are points arbitrarily close that do not belong to F (the point B in fig. 43). All the boundary points taken together form a curve called the *boundary* of the set F. If the set is bounded, then its boundary is represented by a closed curve (see figs. 39, 40).

For what follows, it will be important to notice that *every straight line passing through an interior point of the convex bounded*

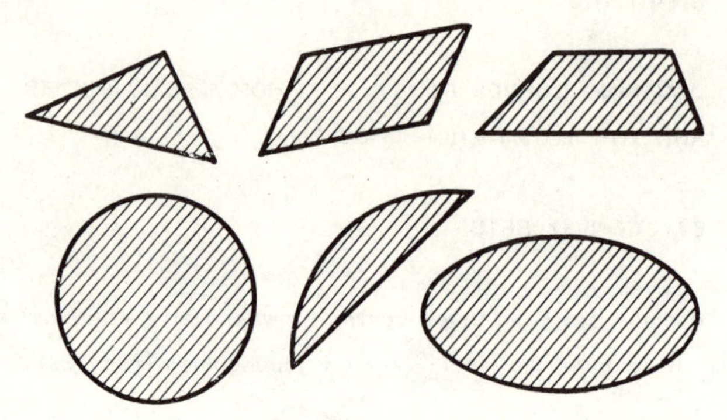

Figure 40.

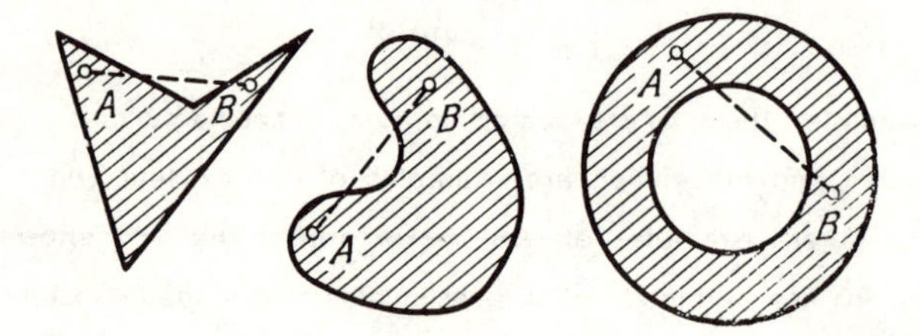

Figure 41.

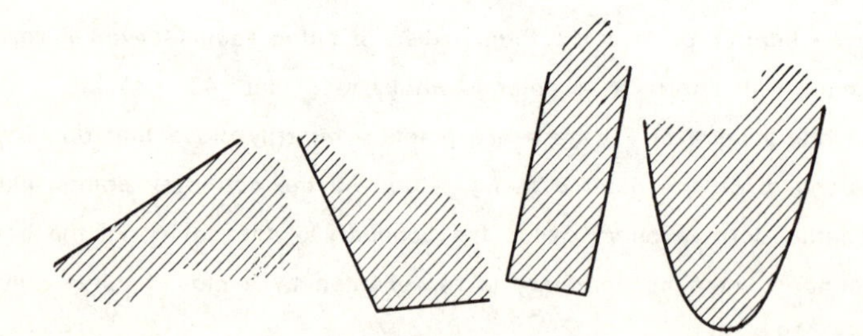

Figure 42.

*set F, cuts the boundary of this set in exactly two points** (fig. 44), moreover, the line segment connecting these two points belongs to *F*, and the entire remaining part of this straight line lies outside *F*.

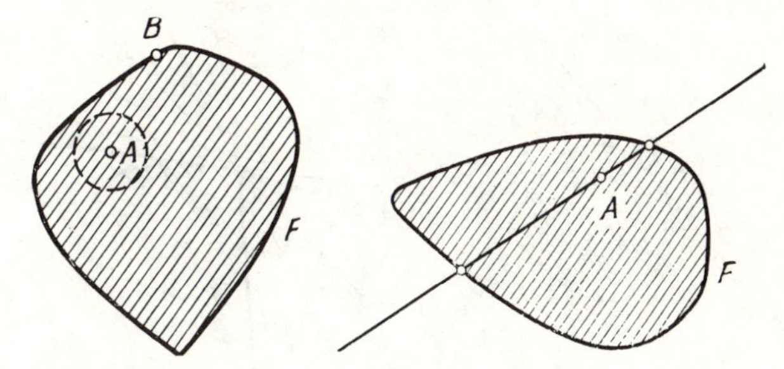

Figure 43. Figure 44.

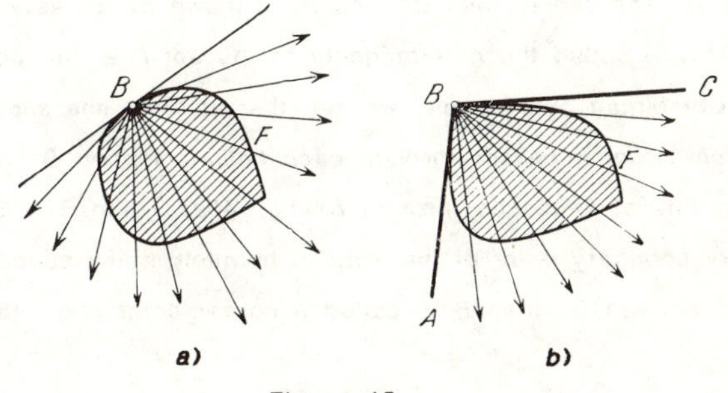

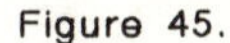

Figure 45.

Let *B* be a boundary point of the convex set *F*. From *B*, draw all possible radial lines passing through points of *F* other than *B*. These radial lines either fill a half-plane (fig. 45*a*) or make up an angle less than 180° (fig. 45*b*). In the first case, the line that bounds the half-plane is a support line of the set *F*. Any other line passing through the point *B* will cut the set into two parts (fig. 46), that is, will not be a support line. In other words, in this case, the unique support line of *F* passes through the point *B*. In the second

---

*The reader may find more detailed information about convex sets (and in particular, the proofs of the properties of these sets mentioned here) in the books [2], [8], [9], [23], [31], [37].

case (fig. 45*b*), the whole set F lies inside the angle ABC which is smaller than 180°, and therefore infinitely many support lines of F pass through B (fig. 47). In particular, the lines BA and BC are

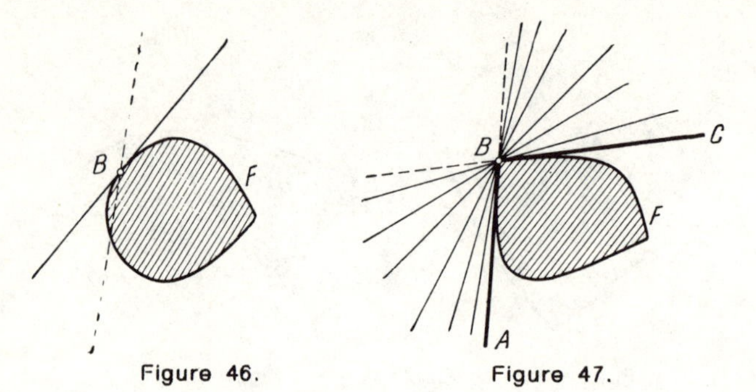

Figure 46. Figure 47.

supports. The radial lines BA and BC (shown by a heavy line in fig. 47) are called the *half-tangents* to the set F at the point B.

Combining both cases, we see that *at least one support line of a convex set F passes through each boundary point B.* If only one support line of F passes through B (fig. 45*a*), then B is called an *ordinary* boundary point of the set. If infinitely many support lines of F pass through B, then B is called a *corner point* (fig. 45*b*).

§8. THE PROBLEM OF COVERING SETS WITH HOMOTHETIC SETS

Let F be a plane set. Choose an arbitrary point O in the plane, and in addition choose a positive number k. For any point A of the set F, we shall find a point A' on the ray OA such that $OA':OA = k$ (fig. 48). The set of all points so obtained is represented by a new set F'. The transition from the set F to the set F' is called *homothety* with centre O and coefficient k, and the set F' itself is called a *homothetic set* of F. (Homothety with a negative coefficient will not be necessary for us in what follows, and we shall therefore not consider it.) If the set F is convex, then its

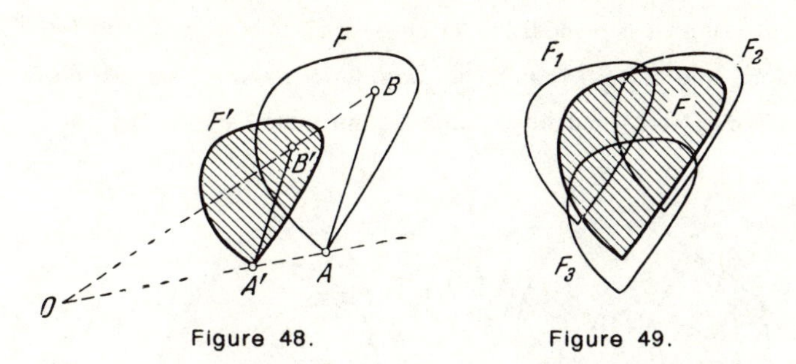

Figure 48. Figure 49.

homothetic set F' is also convex (because if the segment AB belongs wholly to F, then the segment $A'B'$ belongs wholly to F').

Observe that if the coefficient of homothety is less than unity, the set F' (homothetic to F with coefficient k) is represented by a "reduced copy" of the set F.

We now pose the following problem: Given a plane convex bounded set F, find the smallest number of homothetic "reduced copies" of F with which it is possible to cover the whole of F. We shall denote this minimum by $b(F)$. More precisely, the relation $b(F) = m$ means that there exist sets $F_1, F_2, \ldots, F_m$, homothetic to F, with certain centres and coefficients of homothety, the coefficients being less than unity (even if only slightly), such that altogether the sets $F_1, F_2, \ldots, F_m$ cover the whole set F (fig. 49). This number m is minimal, that is, fewer than m homothetic sets are insufficient for this purpose.

It is possible to consider the problem of covering a plane set by smaller homothetic sets not only for plane sets, but also for convex sets in 3-dimensional space (or even in n-space). In 1960 by the Soviet mathematicians I. Ts. Gohberg and A. S. Markus [14] posed the problem of determining the possible values of $b(F)$. Somewhat earlier, this problem (although posed differently) was considered by the German mathematician F. Levi ([29]; see also Problem 14).

For example, consider the case when F is a disc. Then the smaller homothetic sets are arbitrary discs of smaller diameter. It is easy to see that it is impossible to cover the initial disc F with two such discs, that is, $b(F) \geqslant 3$. In fact, let F_1 and F_2 be two discs of smaller diameter, and let O_1 and O_2 be their centres (fig. 50).

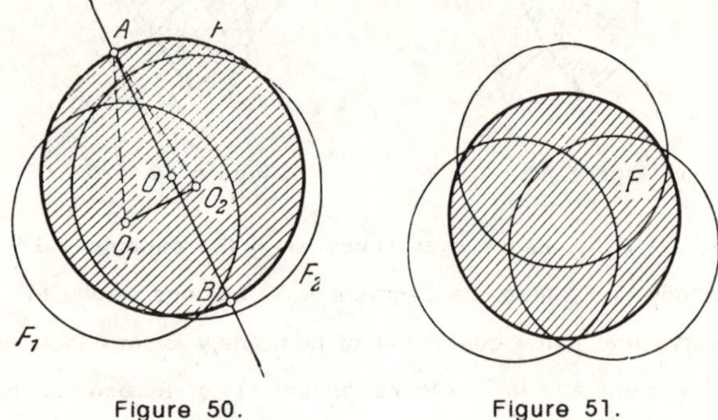

Figure 50. Figure 51.

Draw a perpendicular to the line of the centres O_1O_2 through the centre O of the initial disc F. This perpendicular intersects the circumference of the disc F in two points A and B. Let, for example, the point A lie on the same side of the line O_1O_2 as the point O (if the line O_1O_2 passes through O, then take either of the points A, B). Then $AO_1 \geqslant AO = r$, $AO_2 \geqslant AO = r$, where r is the radius of the disc F. As the discs F_1, F_2 have radii smaller than r, there is one of them to which A does not belong, that is, the discs F_1, F_2 do not cover the whole of the disc F.

On the other hand, it is easy to cover the disc F with three discs of a somewhat smaller diameter (fig. 51). Thus, in the case of the disc, $b(F) = 3$.

Let us now consider the case when F is a parallelogram. It is clear that no parallelogram homothetic to F with coefficient of homothety less than 1 can simultaneously contain two vertices of F. In other words, the four vertices of F must belong to four different homothetic parallelograms, that is $b(F) \geqslant 4$. Four homothetic sets are obviously sufficient (fig. 52). Thus, in the case of the parallelogram, $b(F) = 4$.

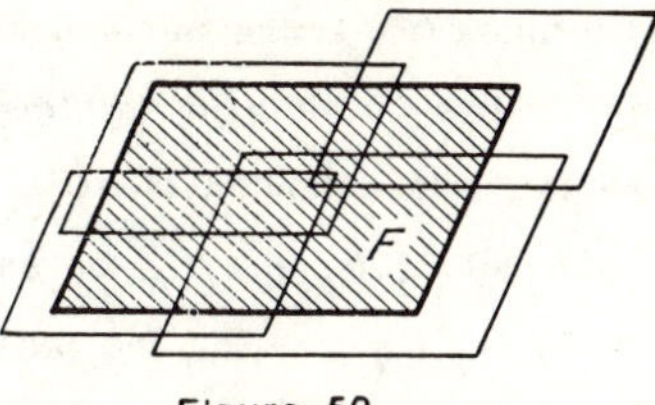

Figure 52.

§9. A REFORMULATION OF THE PROBLEM

Let us reformulate the problem about the covering of a set with smaller homothetic sets in a way resembling Borsuk's problem about the partition of a set into parts of smaller diameter.

Let F be a convex set, and G be one of its parts. We will say that the part G of F *has size equal to k*, if there exists a set F' homothetic to F with coefficient k which contains G, but there is no set homothetic to F with coefficient less than k which contains the whole of G (9). Evidently, if G coincides with all of F, its size equals 1. Therefore, for any part G of F which does not coincide with F, $k \leqslant 1$. However, it should not be supposed that if G does not coincide with the whole of F, then its size is, without fail, less than 1. If, for example, F is represented by a disc, and the part G is an inscribed acute-angled triangle, (fig. 53), then the size of G is equal to 1 (because no disc of smaller diameter can contain the whole of the triangle G). We shall call a part G of the set F a *part of smaller size* if its size $k < 1$.

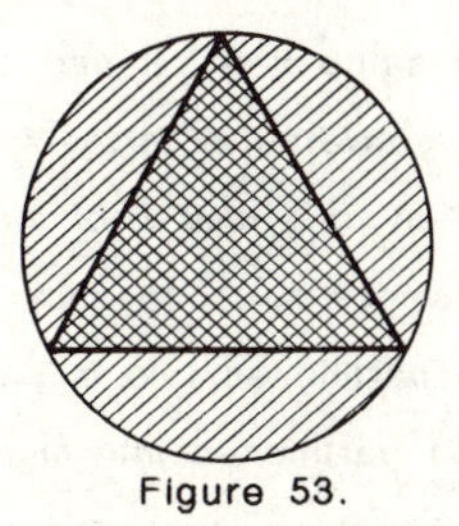
Figure 53.

Making use of the idea of size, we can give the definition of

$b(F)$ in a different form: *$b(F)$ is the minimal number of parts of smaller size into which it is possible to partition the given convex set F.* It is easy to see that this definition of $b(F)$ is equivalent to the previous one. In fact, let $F_1, F_2, \ldots, F_m$ be smaller homothetic sets covering F. Denote by $G_1, G_2, \ldots, G_m$ the parts of the set F being cut out of it by the sets $F_1, F_2, \ldots, F_m$. Clearly, each of the parts $G_1, G_2, \ldots, G_m$ of F has size less than 1. Thus, if the set F may be covered by m smaller homothetic sets, then it is possible to partition it into m parts of smaller size. Conversely, if the set F may be partitioned into m parts $G_1, G_2, \ldots, G_m$ of smaller size, then there exist sets $F_1, F_2, \ldots, F_m$ respectively containing the parts $G_1, G_2, \ldots, G_m$ homothetic to F with coefficients less than unity. These sets $F_1, F_2, \ldots, F_m$ form a cover of F by smaller homothetic parts.

It is clear that all the above (that is, the definition of size and the other definition of $b(F)$) applies not only to planar sets, but also to convex sets of any number of dimensions. Thus, the problem of covering a convex set with smaller homothetic sets may be stated as *the problem of partitioning a set F into parts of smaller size*. In this form, it very much resembles Borsuk's problem studied in Chapter 1.

However, the connection between these problems is not purely superficial. In fact, if the set F has diameter d, then the set homothetic to F with coefficient k has diameter kd. From this, it follows that if a convex set F has diameter d, then each of its parts of smaller size is, at the same time, a part of smaller diameter. (Generally speaking, the converse is false; for example, an equilateral triangle inscribed in a disc F of diameter d is a part of smaller diameter, but has size equal to unity; fig. 53.) Therefore, if a convex set F can be partitioned into m parts of smaller size, then, a fortiori, it may be partitioned into m parts of smaller diameter (but, generally speaking, the converse is false, as shown by the example of the parallelogram).

Thus, for any convex set F, we have the inequality:

$$a(F) \leqslant b(F). \qquad (*)$$

Besides plane sets, this assertion is true for convex bodies of any number of dimensions (see Problem 7).

Note that the problem of the partition into parts of smaller size depends on the convexity of the sets, whereas Borsuk's problem about a partition into parts of smaller diameter is posed for any (even non-convex) set. However, this is immaterial; it is easily seen that if Borsuk's conjecture were confirmed for n-dimensional convex sets, its validity would follow for any n-dimensional set. In fact, for any set F of diameter d, there exists a smallest convex set $\tilde{F}$ containing it; this convex set (fig. 54), called the *convex hull* of

Figure 54.

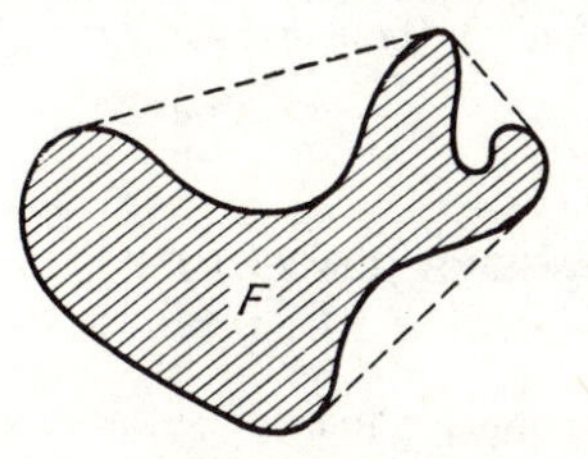

F, has the same diameter d. From this it follows that to determine the possibility of a partition into $n+1$ parts of smaller diameter, it is sufficient to consider only convex n-dimensional sets.

§10. SOLUTION OF THE PROBLEM FOR PLANE SETS

As we saw in §8, in the problem of covering a convex set with smaller homothetic sets (as opposed to Borsuk's problem), the disc is not a set which requires the greatest number of covering sets. $b(F)$ is greater for the parallelogram than for the disc.

The question naturally arises as to whether there exist plane convex sets for which $b(F)$ is even greater than for the parallelogram. It turns out that such sets do not exist; furthermore, among all plane convex sets, the equality $b(F) = 4$ is realized only for parallelograms. In other words, we have the following theorem established in 1960 by I. Ts. Gohberg and A. S. Markus [14] (somewhat earlier, in 1955, F. Levi [30] obtained another result essentially coinciding with this theorem; see Problem 14):

Theorem 5. *If F is a plane bounded convex set other than a parallelogram, then $b(F) = 3$; if F is a parallelogram, then $b(F) = 4$.*

We shall not present the proof of this theorem immediately, as we will obtain, in §14, this theorem as a consequence of other results. We notice only that Theorem 5 gives a new proof of Theorem 1. In fact, if the plane set F is not a parallelogram, then, by virtue of Theorem 5, $b(F) = 3$, and therefore, $a(F) \leqslant 3$ (see inequality (*) above). If F is a parallelogram, then $a(F) = 2$ (Figure 12*b*). Thus, in both cases, $a(F) \leqslant 3$.

§11. HADWIGER'S CONJECTURE

After solving the problem of covering plane sets with smaller homothetic sets, it is natural to turn our attention to this problem for spatial bodies. For what 3-dimensional body F does $b(F)$ take its maximum value? Based on the theorem stated in the previous section, it is natural to conjecture that a parallelepiped is such a convex body in 3-space. As is easily seen, for a parallelepiped F we have $b(F) = 8$. In fact, no parallelepiped homothetic to F with coefficient of homothety less than 1 can simultaneously contain two vertices of F. Consequently, the eight vertices of F must belong to different homothetic parallelepipeds, that is, $b(F) \geqslant 8$. Eight homothetic parallelepipeds are obviously sufficient; for example, it is possible to partition F into 8 homothetic parallelepipeds (with coefficient $k = 1/2$), obtained by drawing three planes parallel to the faces of F through the centre of F.

An analogous situation exists for an *n-dimensional*

parallelepiped F for which $b(F) = 2^n$ for any n.

Is this value of $b(F)$ maximal? In other words, *can any n-dimensional convex body F be partitioned into* 2^n *parts of smaller size* (or, equivalently, may be covered by 2^n smaller homothetic bodies)? If so, then *are the n-dimensional parallelepipeds the unique convex bodies for which* $b(F) = 2^n$? In 1957, Hadwiger [21] published a list of unsolved geometric problems. Among them were both the above problems. There, he conjectured that both problems have positive solutions, that is, that $b(F) \leqslant 2^n$ for any bounded convex n-dimensional body, and equality is achieved only in the case of the parallelepiped. This conjecture was independently posed by I. Ts. Gohberg and A. S. Markus [14].

These problems have not yet been solved. Their solution is not even known for $n = 3$. In other words, it is not known whether *any three-dimensional convex body may be partitioned into eight parts of smaller size* (or may be covered by 8 smaller homothetic bodies). Furthermore, the solutions of these problems are not even known for n-dimensional polytopes (see Problem 8).

However, the Soviet mathematicians A. Yu. Levin and Yu. I. Petunin proved that *for any n-dimensional centrally symmetric convex body F*, $b(F) \leqslant (n+1)^n$. For three-dimensional convex bodies, this means that $b(F) \leqslant 4^3 = 64$. As we see, this bound is very far from Hadwiger's conjecture. Finally, Rogers (see [18]) obtained the following bound for centrally symmetric bodies:

$$b(F) \leqslant 2^n(n \ln n + n \ln \ln n + 5n)$$

Hadwiger's conjecture gives the expected upper bound for $b(F)$. It is possible to determine the lower bound for $b(F)$ exactly: *for any n-dimensional bounded convex body, the inequality* $b(F) \geqslant n+1$ *holds*; in particular, for a three-dimensional convex body, $b(F) \geqslant 4$. In addition, there exist bodies (for example, the n-dimensional ball), for which $b(F) = n+1$. We shall give the proof of the inequality $b(F) \geqslant n+1$ below (see §15).

Notice also that for any m satisfying the inequalities $4 \leqslant m \leqslant 8$, there exists a convex body (and even a polytope) in 3-space for which $b(F) = m$. These polytopes are obtained from a cube cut off at certain vertices (fig. 55).

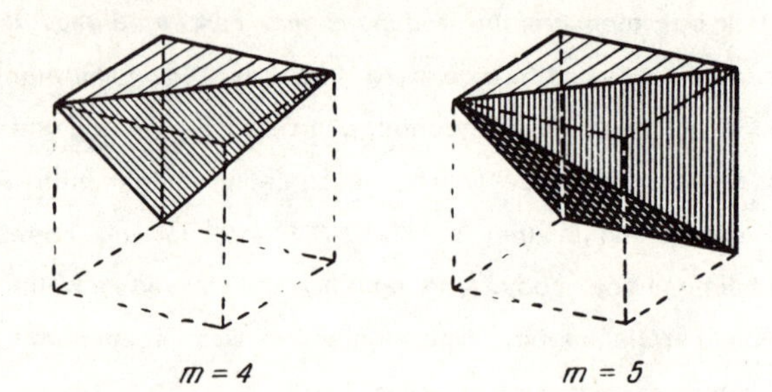

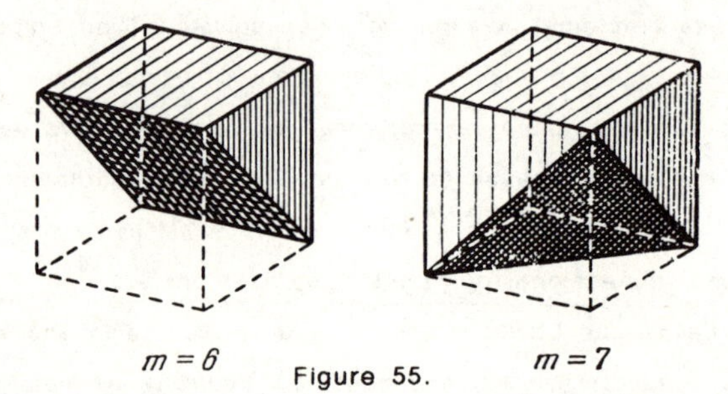

Figure 55.

An analogous situation holds in n-dimensional space: for any m satisfying $n + 1 \leqslant m \leqslant 2^n$, there exists a bounded convex body (and even a polytope) in n-dimensional space, for which $b(F) = m$.

§12. THE ILLUMINATION PROBLEM

Let F be a plane bounded convex set, and ℓ be an arbitrary direction in the plane of this set. We shall say that a boundary point A is a point of illumination with respect to the direction ℓ if the parallel beam of rays having direction ℓ "illuminates" the point A on the boundary of the set F and some neighbourhood of A (fig. 56). Notice that if the line parallel to ℓ that passes through A is a

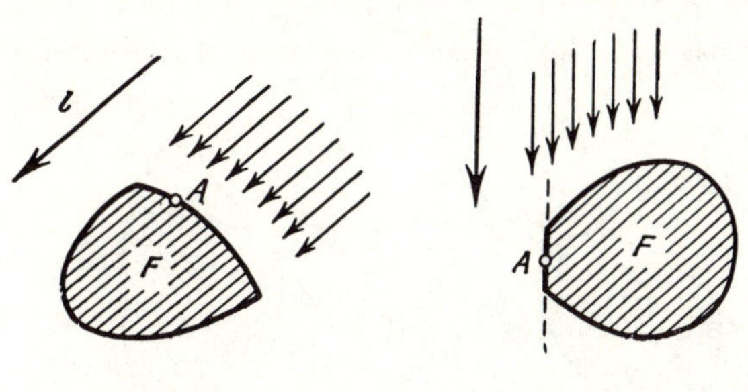

Figure 56. Figure 57.

support of the set F (fig. 57), then we do not consider the point A as a point of illumination with respect to ℓ. In other words, the point A is a point of illumination if it satisfies the following two conditions:

1). The line p, parallel to ℓ and passing through A is not a support line of the set F (that is, interior points of F lie on p).

2). A is the first point of F which we meet moving along p in the direction ℓ.

Let us agree to say that the directions $\ell_1, \ell_2, \ldots, \ell_m$ taken together are *sufficient to illuminate the boundary* of F if each boundary point of F is a point of illumination with respect to at least one of these directions. Lastly, let us denote by $c(F)$ the smallest natural number m such that there exist m directions in the plane which together are sufficient to illuminate the whole boundary of F. It is possible to consider the problem of determining $c(F)$ or, as we shall call it, the problem of illuminating the boundary of F not only for plane sets, but also for convex sets in 3-space (or even for n-space). The points of illumination are defined by the same conditions (1) and (2) as in the case of plane sets (fig. 58). The illumination problem was posed in 1960 by the Soviet mathematician V.G. Boltyansky [1].

It is easy to prove that $c(F)$ is always at least 3 for any plane set F. In fact, let F be a bounded convex set in the plane, and ℓ_1, ℓ_2 be arbitrary directions. Draw two support lines of F parallel to ℓ_1, and let A and B be boundary points of F lying on these support planes (fig. 59). Then neither A nor B is a point of

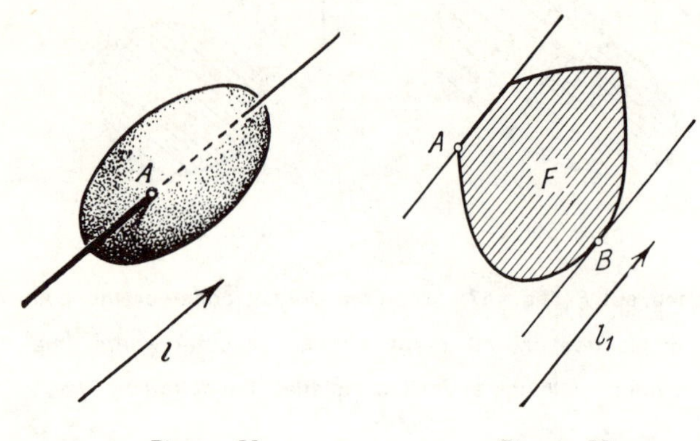

Figure 58. Figure 59.

illumination for the direction ℓ_1, and the direction ℓ_2 may illuminate at most one of these points. Hence two directions are not sufficient to illuminate the whole boundary of F.

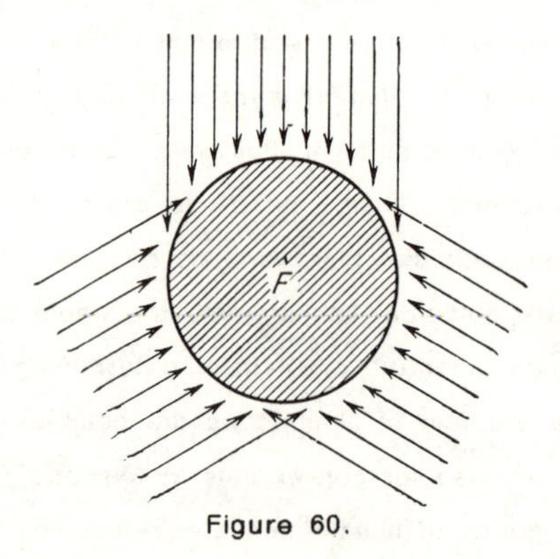

Figure 60.

In the case of the disc (fig. 60), three directions are sufficient to illuminate the boundary. For the parallelogram (fig. 61), three directions are insufficient (because no direction can simultaneously illuminate two vertices), but four directions permit the illumination of the whole boundary of the parallelogram. In other words, for the disc $c(F) = 3$, and for the parallelogram, $c(F) = 4$.

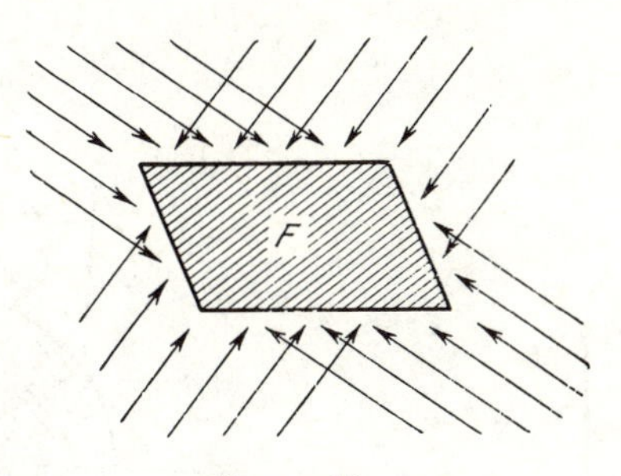

Figure 61.

§13. A SOLUTION OF THE ILLUMINATION PROBLEM FOR PLANE SETS

As in the case of the problem of covering a set with smaller homothetic parts, the parallelogram plays a special role in the illumination problem. Namely, we have the following:

Theorem 6. *For any bounded plane set F other than a parallelogram, c(F) = 3; if F is a parallelogram, then c(F) = 4.*

Proof. Firstly, let us suppose that the set *F* does not have a corner point. In this case, we shall choose three directions ℓ_1, ℓ_2, ℓ_3 subtending angles of 120° with each other (fig. 62), and show that these three directions illuminate the whole boundary of *F*. In fact, let *A* be an arbitrary boundary point of *F* (fig. 63). Let us draw a support line *p* of *F* passing through *A*. Furthermore, draw three vectors beginning at *A* and having directions ℓ_1, ℓ_2, ℓ_3; we shall denote the ends of these vectors by *K*, *L*, *M* respectively. Then *A* lies inside the triangle *KLM*. As the line *p* passes through the interior point *A* of the triangle *KLM*, it partitions this triangle into two parts. From here it follows that both sides of the line *p* contain vertices of the triangle *KLM*. Choose a vertex of the triangle *KLM* lying on the same side of *p* as the set *F*; let this be, for example, the vertex *M* (corresponding to the direction ℓ_3). The line *AM* is

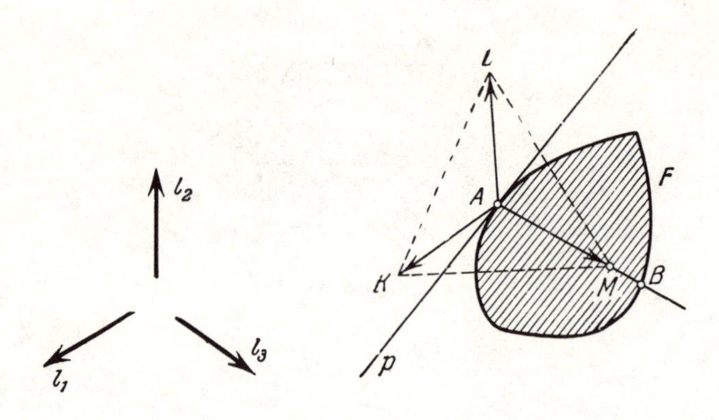

Figure 62. Figure 63.

not a support of F (because it is different from the line p, and F does not have corner points, and therefore has a unique support line at each boundary point). In other words, the line AM partitions the set F, that is, passes through interior points of this set. The line AM intersects the boundary of F in two points, one of which is A; we shall denote the second point of intersection by B. The points B and M lie on one side of A. Consequently, if we move along the line AM in the direction ℓ_3, then A will be the first point of F which we meet. Furthermore, as the line AM passes through interior points of F, the *direction of* L_3 *illuminates the point* A. And so, whichever boundary point of F we chose A to be, it is a point of illumination for at least one of the directions ℓ_1, ℓ_2, ℓ_3, and therefore, $c(F) = 3$.

Now let us suppose that F has corners, and let A be one of them. Draw two half-tangents to F at A, and also draw two support lines parallel to these half-tangents (fig. 64). We get a parallelogram $ABCD$ around F. Firstly, consider the case when the vertex C of this parallelogram does not belong to F. We shall denote by M and N the points of F nearest to C lying on the sides CB and CD. The points M and N partition the boundary of F into

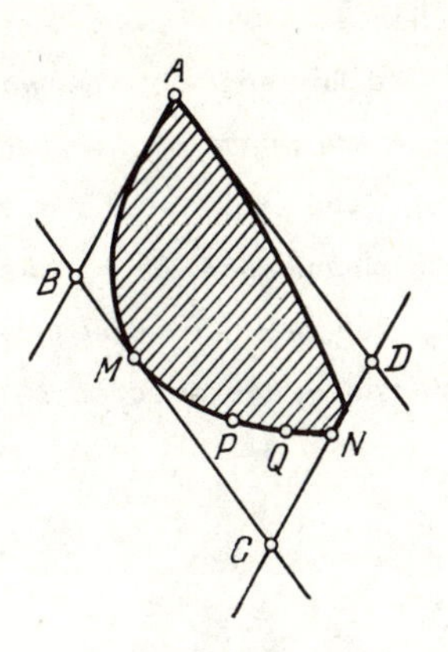

Figure 64.

two parts; we shall consider the one which does not contain A, and take two points P and Q on this arc chosen so that the points lie on the boundary of F in the following order: M, P, Q, N. Notice that P and Q lie inside the parallelogram $ABCD$.

We now show that the directions $\ell_1 = \overline{QD}$, $\ell_2 = \overline{PB}$ and $\ell_3 = \overline{AC}$ illuminate all the boundary points of F. In fact, the line $\overline{QD}$ intersects the boundary of F in two points, one of which is Q (fig. 65). From this, it is easy to deduce that Q is a point of

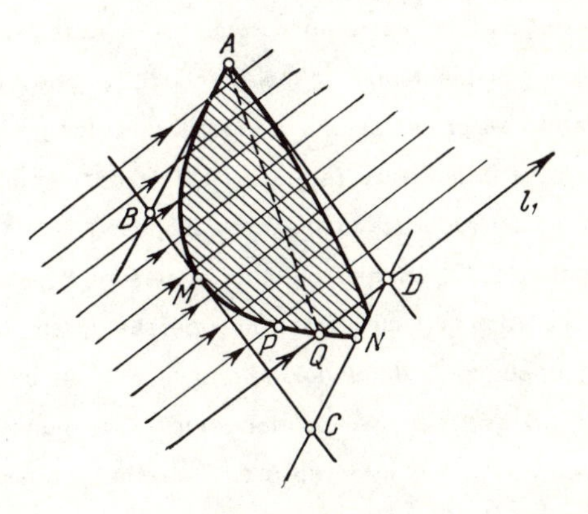

Figure 65.

illumination for the direction ℓ_1. By drawing lines parallel to QD through all the points of the line segment AQ, we find that *all points of the arc QA apart from A are points of illumination for the direction* ℓ_1. Analogously, (fig. 66), all the points of the arc AP apart from A are points of illumination for the direction ℓ_2. Thus, the directions ℓ_1 and ℓ_2 illuminate all the boundary of F apart from A. The point A is illuminated by the direction ℓ_3.

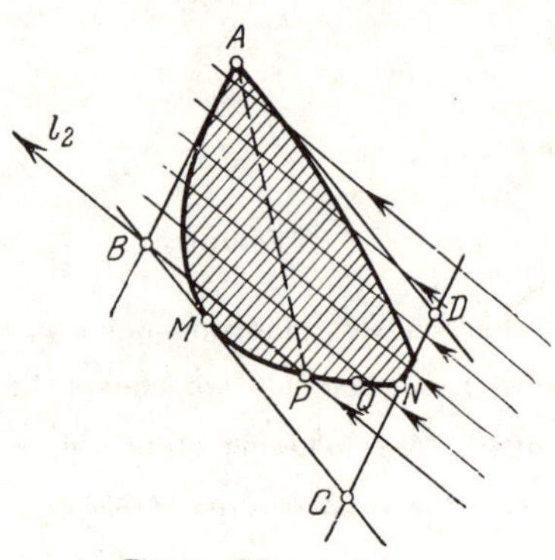

Figure 66.

So the directions ℓ_1, ℓ_2, ℓ_3 illuminate the whole boundary of the set F, that is, $c(F) = 3$.

We considered the case when the vertex C of the parallelogram $ABCD$ does not belong to F. The next case is when C belongs to F, but at least one of the points B, D does not. Without loss of generality, we can assume B does not belong to F. The case when at least one of the rays $\overline{CB}$, $\overline{CD}$ is not a half-tangent at the point C is easily reduced to the previous case. It is sufficient to draw half-tangents CM, CN and support planes of F parallel to them (fig. 67); in the resulting circumscribing parallelogram $CB'A'D'$, the vertex A' lying opposite C does not belong to F. Let us suppose, therefore, that $\overline{CB}$ and $\overline{CD}$ are half-tangents. The points A and C partition the boundary of F into two arcs; we shall consider the arc which lies on the same side of the diagonal AC as the point B. Let us take two points P and Q on this arc lying inside the parallelogram

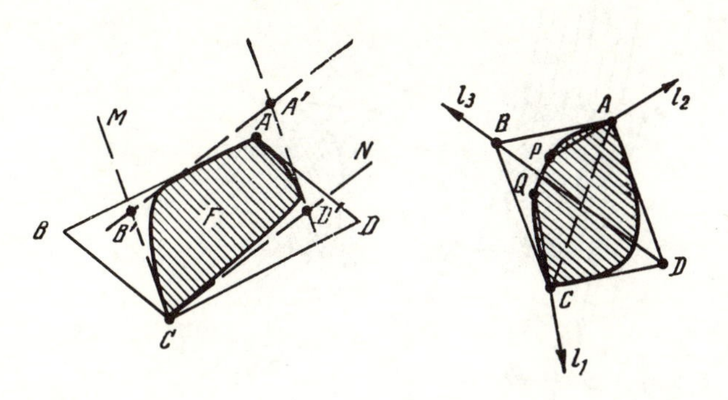

Figure 67. Figure 68.

ABCD (fig. 68). In addition, we choose to label *P* and *Q* such that the points lie on the boundary of *F* in the order *A, P, Q, C*. We shall show that, in this case, the directions:

$$\ell_1 = \overline{QC}, \qquad \ell_2 = \overline{PA}, \qquad \ell_3 = \overline{DB}$$

illuminate the whole boundary of *F*. In fact, drawing lines parallel to QC through all the points of the segment *QA* (fig. 69*a*), we find that *all points of the arc QA are points of illumination for the direction* ℓ_1. In particular, *A* is a point of illumination for ℓ_1. In fact, the line passing through *A* parallel to *QC* goes inside the parallelogram, and therefore must pass through interior points of *F* (because *AB* and *AD* are half-tangents).

So the direction ℓ_1 illuminates all points of the arc *AQ*, including *A*. Analogously, the direction ℓ_2 illuminates all points of the arc *PC*, including *C*. Together, the directions ℓ_1 and ℓ_2 illuminate all points of the arc *APQC*, including *A* and *C* (fig. 68). The remaining points are illuminated by the direction ℓ_3 (fig. 69*b*). Thus, $c(F) = 3$.

We have remaining only the unsolved case when all four vertices of the parallelogram *ABCD* belong to *F*. But in this case, by virtue of convexity, the set *F* coincides with the parallelogram *ABCD*, and therefore $c(F) = 4$, completing the proof of Theorem 6.

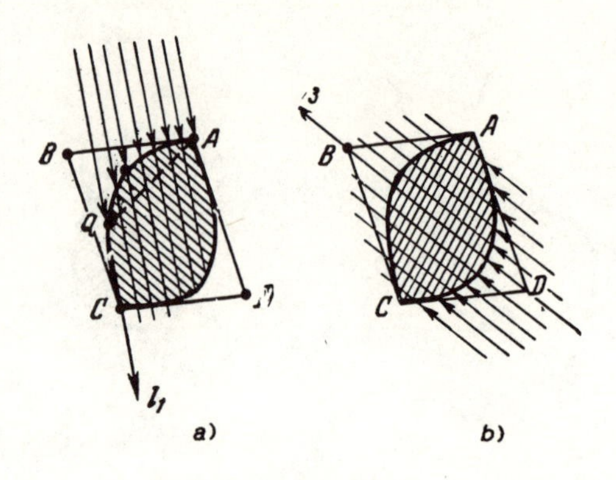

Figure 69.

§14. THE EQUIVALENCE OF THE TWO PROBLEMS

The reader, no doubt, will have already observed that for the disc and the parallelogram, $b(F)$ and $c(F)$ are the same. It is also striking that the statements of Theorems 5 and 6 are absolutely identical, except for the substitution of $c(F)$ for $b(F)$.

In other words, for plane bounded convex sets, the values of $b(F)$ and $c(F)$ coincide. This holds not only for plane sets, but also for convex bounded sets of any number of dimensions. In other words, we have the following theorem proved in 1960 by V. G. Boltyansky [1]:

Theorem 7. *If F is an n-dimensional bounded convex body, then:*

$$b(F) = c(F). \tag{**}$$

This theorem means that the illumination problem is equivalent to the problem of covering a convex body with smaller homothetic bodies. Moreover, the illumination problem clearly has the advantage of being easy to visualize. Notice that Theorem 7 immediately implies Theorem 5 which we have not yet proved. In fact, on the strength of the equality (**), Theorems 5 and 6 directly follow from each other, and Theorem 6 has already been proved. (The original proof of Theorem 5, given by I. Ts. Gohberg and A. S. Markus, without using of the equality (**) was more complicated than the proof of Theorem 6.) Notice further that equality (**) allows us to reformulate Hadwiger's conjecture from first principles: *is it true that the boundary of any n-dimensional bounded convex body F may be illuminated by 2^n directions, and that, if F is not an n-dimensional parallelepiped, then $2^n - 1$ directions are sufficient?*

We have already noted that the truth of this conjecture has not yet been established for $n = 3$, that is, up to now, we have not even proved that *the boundary of any bounded convex body in three-dimensional space may be illuminated by eight directions.* This has not been proved even for convex polytopes (see Problem 9).

Let us go through the proof of Theorem 7 using plane convex sets as an example. For convex bodies (of any number of dimensions), the proof works in an analogous manner, but with some complications which we shall mention in the notes.

Let us suppose that it is possible to cover a plane convex set F with smaller homothetic sets $F_1, F_2, \ldots, F_m$. Denote the centre of homothety corresponding to F_i by O_i, and the coefficient of this homothety by k_i $(i = 1, 2, \ldots, m)$. Thus, each of the numbers $k_1, k_2, \ldots, k_m$ is less than 1.

Now choose an arbitrary interior point A of F, not coinciding with any of the points $O_1, O_2, \ldots, O_m$ and denote by $\ell_1, \ell_2, \ldots, \ell_m$ the directions defined by the rays:

$$\overline{O_1A}, \overline{O_2A}, \ldots, \overline{O_mA}.$$

We shall now show that the directions $\ell_1, \ell_2, \ldots, \ell_m$ are together sufficient to illuminate of the whole boundary of F. In fact, let B be an arbitrary boundary point of F (fig. 70). Then B belongs to at least one of the sets $F_1, F_2, \ldots, F_m$, say, for example, the set F_i. As F is mapped into F_i by the homothety with centre O_i and

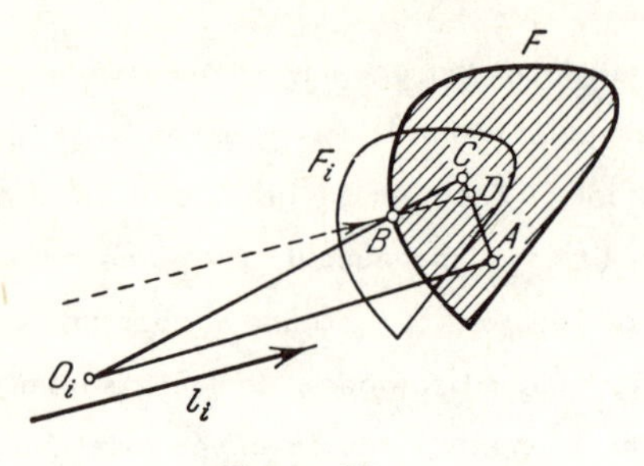

Figure 70.

coefficient k_i, then there is a point C of F which is mapped by the homothety to B. Thus, $O_iB:O_iC = k_i$. The equality $O_iB:O_iC = AD:AC$ implies that $BD \parallel O_iA$, that is the line BD is parallel to the direction ℓ_i. Furthermore, as the point C belongs to F and A is an interior point of this set, all the points of the line segment AC (except, perhaps, C) are interior points of F; in particular, D is an interior point of this set.

So the line BD is parallel to the direction ℓ_i and passes through the interior point D of F. From this, it follows that B is a point of illumination with respect to the direction ℓ_i. Thus, any boundary point of F is illuminated by one of the directions $\ell_1, \ell_2, \ldots, \ell_m$.

We have proved that if the set F may be covered by m smaller homothetic sets, then m directions are sufficient for the illumination of its boundary. Consequently, we have the inequality:

$$c(F) \leqslant b(F)$$

We shall now establish the opposite inequality:

$$c(F) \geqslant c(F)$$

Suppose that s directions $\ell_1', \ell_2', \ldots, \ell_s'$ are together sufficient to illuminate all the boundary of F. Draw two support lines of F parallel to the direction ℓ_i' (fig. 71), and denote by A and B the first points we meet moving along this line in the direction ℓ_i'. Then it is clear that all the points of the arc Δ_i with ends A, B (shown by

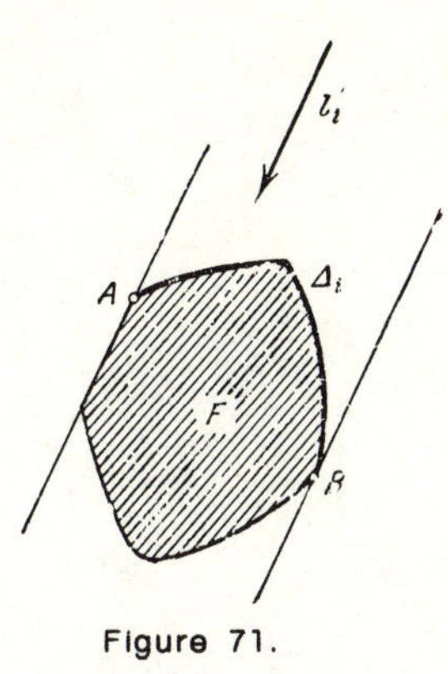

Figure 71.

the thick line in fig. 71), apart from the endpoints A and B, are points of illumination with respect to the direction ℓ_i'. Thus, the set of all illuminated points with respect to the direction ℓ_i' is represented by an arc Δ_i without the endpoints. We shall call this set the *region of illumination* with respect to the direction ℓ_i' (10). As the directions $\ell_1', \ell_2', \ldots, \ell_s'$ illuminate the whole boundary of F, the corresponding regions of illumination $\Delta_1, \Delta_2, \ldots, \Delta_s$ cover all the boundary of F.

The point A in Figure 71 is not a point of illumination with respect to the direction ℓ_i', and therefore is illuminated by one of the other directions $\ell_1', \ell_2', \ldots \ell_s'$, for example, by the direction ℓ_j'. But then the direction ℓ_j' illuminates all points sufficiently close to A, that is, the regions of illumination Δ_i and Δ_j overlap (fig. 72). In just the same way, the end B of the arc Δ_i overlaps another region of illumination Δ_k.

The fact that the arcs $\Delta_1, \Delta_2, \ldots, \Delta_s$ are regions of illumination, with the ends overlapping one another, implies that we can slightly reduce them, and these reduced arcs will still cover the whole boundary of F. In other words, it is possible to choose arcs $\Delta_1^*, \Delta_2^*, \ldots, \Delta_s^*$ contained (together with their endpoints) inside $\Delta_1, \Delta_2, \ldots, \Delta_s$ (fig. 73), such that together, the arcs $\Delta_1^*, \Delta_2^*, \ldots, \Delta_s^*$ cover the whole boundary of F (11).

We denote the ends of the arc Δ_i by A and B, and the ends

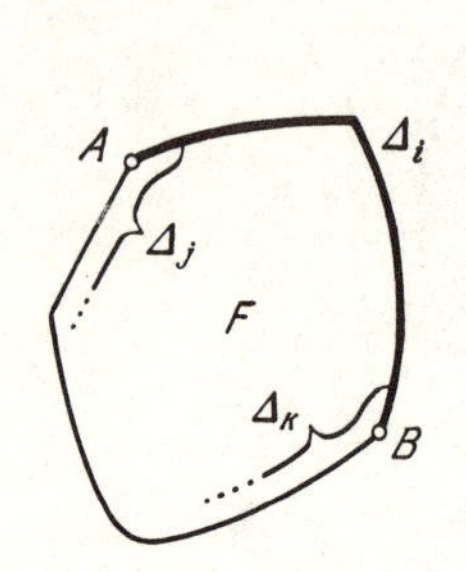

Figure 72.

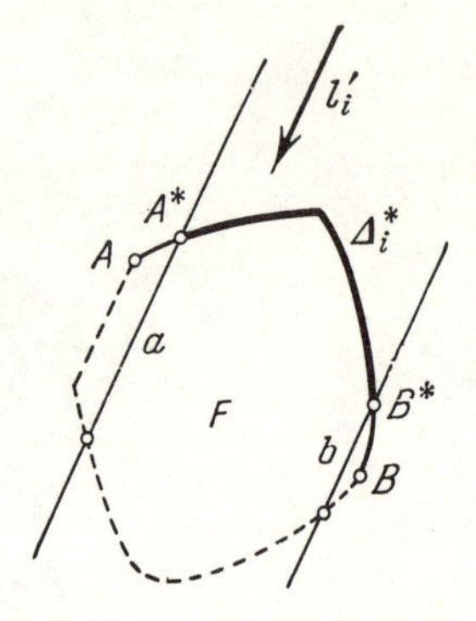

Figure 73.

of the arc Δ_i^* by A^* and B^*. The lines passing through the points A^* and B^* parallel to the direction ℓ_i' must pass through interior points of F (because A^* and B^* are points of illumination with respect to the direction ℓ_i'). We shall denote by a and b the lengths of the chords on these lines being cut by F, and shall choose segments h_i' less than a and b. Then the parallelogram having one side as A^*B^*, and the second side parallel to ℓ_i' with length h_i, lies wholly inside the set F (fig. 74). From here, it

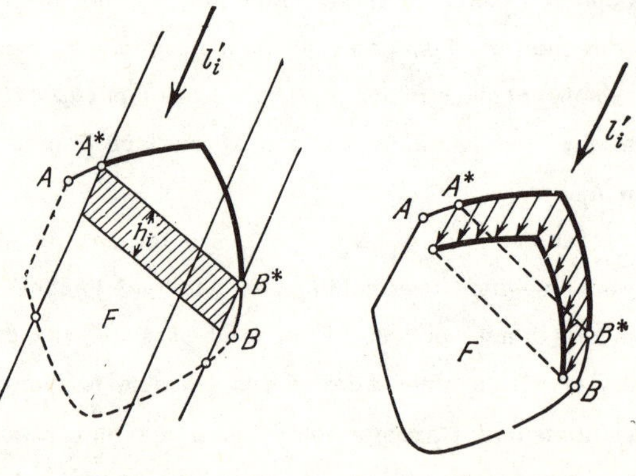

Figure 74. Figure 75.

follows that the set F cuts a segment of length greater than h_i from any line parallel to the direction ℓ_i', and passing through some point of the arc Δ_i^*. This means that a parallel translation of the arc Δ_i^* in the direction ℓ_i' by a distance h_i (fig. 75) moves the arc Δ_i^* wholly inside F (12). In other words, performing a parallel translation of F in the direction opposite to ℓ_i' by a distance h_i, we obtain a set F_i^* whose *interior* contains the arc Δ_i^* (fig. 76). Therefore, choosing an arbitrary point O_i^* inside F_i^* and producing the homothety of F_i^* with centre O_i^*, and coefficient $k_i^* < 1$ sufficiently close to unity, we obtain a set F_i' homothetic to F_i^* (which means also to F), and containing the arc Δ_i^*. We carry out this construction for all $i = 1, 2, \ldots, s$, and obtain sets $F_1', F_2', \ldots, F_s'$ homothetic to F with coefficient of homothety less than unity.

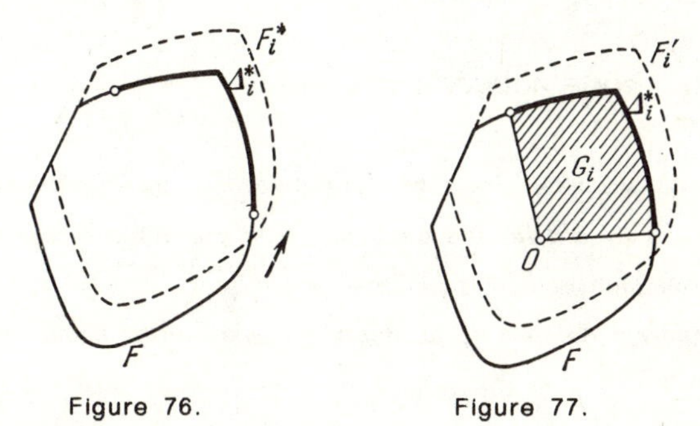

Figure 76. Figure 77.

Now let O be some interior point of F. We may suppose that the above constructions are carried out so that each set $F_1', F_2', \ldots, F_s'$ contains O (fig. 77). For this, it is sufficient to take the segment h_i sufficiently small, and the coefficients k_i^* sufficiently close to unity.

Lastly, let us denote by G_i the "sector" with apex O and arc Δ_i^* (this sector is shaded in fig. 77). As the set F_i' is convex and, furthermore, contains the arc Δ_i^* and the point Δ, then F_i' contains the whole sector G_i. Consequently, the sets $F_1', F_2', \ldots, F_s'$ together contain all the sectors $G_1, G_2, \ldots, G_s$. But it is clear that the sectors $G_1, G_2, \ldots, G_s$ cover all of F (because the arcs

$\Delta_1^*, \Delta_2^*, \ldots, \Delta_s^*$ cover all the boundary of F). Therefore, the sets $F_1', F_2', \ldots, F_s'$ cover all of F (13).

We have proved that if all the boundary of F may be illuminated by s directions, then F may be covered by s smaller homothetic sets. Consequently, we have the inequality:

$$b(F) \leqslant c(F)$$

The inequalities $c(F) \leqslant b(F)$ and $b(F) \leqslant c(F)$ which we have proved, imply the equality:

$$b(F) = c(F)$$

completing the proof of Theorem 7.

§15. SOME BOUNDS FOR $c(F)$

Here, we shall prove two straightforward theorems which, in particular, fully answer the question about the value of $c(F)$ for convex sets with smooth boundary.

Theorem 8. *If F is an n-dimensional convex body, then $c(F) \geqslant n + 1$.*

Proof. We shall go through the proof for three-dimensional convex bodies ($n = 3$); for other values of n, the proof is completely analogous. Let F be an arbitrary three-dimensional convex body, and ℓ_1, ℓ_2, ℓ_3 be three arbitrary directions. We shall show that these directions cannot illuminate the whole boundary of F. The rays ℓ_1, ℓ_2, ℓ_3 will be considered as originating from a single point O. Draw the plane Γ, passing through the rays ℓ_1, ℓ_2. We shall suppose that this plane is "horizontal", and that of the two half-spaces defined by this plane, the one in which the last ray ℓ_3 lies is the "upper" half-space (fig. 78*a*). (If all three rays lie in one plane, then it is possible to consider either of the two half-spaces as being "upper".) Draw now a horizontal support

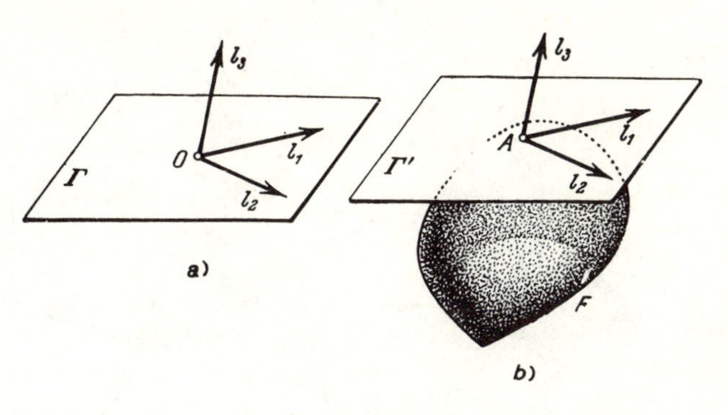

Figure 78.

hyperplane Γ' to F (that is, parallel to Γ) with respect to which the body F lies in the lower half-space (fig. 78*b*), and let A be any common point of the plane Γ' and the body F. Then the point A is not illuminated by any of the directions ℓ_1, ℓ_2, ℓ_3; the rays ℓ_1, ℓ_2, lying in the support plane Γ' obviously do not illuminate A, and also the ray ℓ_3 does not illuminate this point because it comes out from A into the upper half-space, whereas F lies in the lower half-space. Hence three directions are not enough to illuminate the boundary of F, and so $c(F) \geqslant 4$.

Theorem 9. *If F is a convex n-dimensional body with smooth boundary, then n+1 directions are sufficient to illuminate its boundary, that is* $c(F) = n + 1$.

We have already deduced the proof of this theorem for $n = 2$ (see the beginning of the proof of Theorem 6). For arbitrary n, the proof is analogous. Namely, take an arbitrary n-dimensional simplex (that is, an "n-dimensional tetrahedron"), and from its interior point O draw $n+1$ rays to its vertices $B_1, B_2, \ldots B_{n+1}$ (fig. 79). This will give us $n+1$ directions $\ell_1, \ell_2, \ldots, \ell_{n+1}$, sufficient for the illumination of the boundary of an n-dimensional convex body F with smooth boundary. In fact, let A be an arbitrary boundary point of F and Γ be a support hyperplane of F passing through this point. Construct a parallel translation of the simplex with vertices $B_1, B_2, \ldots, B_{n+1}$ mapping the point O to A (fig. 80). Denote the

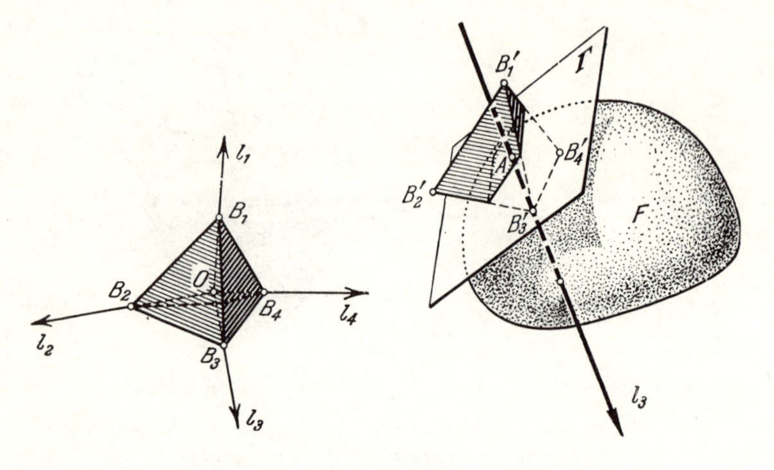

Figure 79. Figure 80.

vertices of the simplex in the new position by $B_1', B_2', \ldots, B_{n+1}'$. As the hyperplane Γ passes through the interior point A of the simplex $B_1', B_2', \ldots, B_{n+1}'$, it cuts this simplex, that is points of this simplex lie on both sides of Γ. Let B_i' be a vertex lying on the same side of Γ as the body F (vertex B_3' in fig. 80). The line AB_i' is not a support for the body F (because F does not have corner points, and therefore, all its support lines passing through A lie in the hyperplane Γ). In other words, the line AB_i' (parallel to OB_i, that is, having direction ℓ_i) *passes through interior points of F*. From this, it is easily deduced that the direction ℓ_i illuminates A. Thus, the directions $\ell_1, \ell_2, \ldots, \ell_{n+1}$ illuminate the whole boundary of F, that is, $c(F) \leqslant n + 1$. But the opposite inequality is given by Theorem 8, and so $c(F) = n + 1$.

Remark. By a slightly more involved proof, it is possible to arrive at the following theorem, proved in 1960 by V. G. Boltyansky [1]: *if an n-dimensional convex body has no more than n corners, then* $c(F) = n + 1$ (14). (See Problem 10 in connection with this)

Corollary. *If F is an n-dimensional convex body then:*

$b(F) \geqslant n + 1$

If F has smooth boundary (or has no more than n corner points), then:

$$a(F) \leqslant b(F) = n + 1$$

Thus, for an n-dimensional convex body with smooth boundary (and even for an n-dimensional convex body having no more than n corner points), Borsuk's conjecture is true.

This follows immediately from the relations (*) and (**) above. Thus, we have obtained here a new proof of Hadwiger's theorem (Theorem 4), and even a somewhat stronger result.

§16. PARTITION AND ILLUMINATION OF UNBOUNDED CONVEX SETS

We shall state here results due mainly to the Soviet mathematician P.S. Soltan. We shall not as a rule give the proofs here, refering the reader to Soltan's original article [35].

For unbounded convex sets (see fig. 42), Borsuk's problem is undefined, as the diameter of the sets becomes infinite. However, the illumination problem and the problem of covering sets with smaller homothetic sets (that is, sets homothetic to the given set with coefficient of homothety less than unity) retains its meaning as before. Here a surprise awaits us: *Theorem 7 about the equality of $b(F)$ and $c(F)$ no longer holds for unbounded convex sets.*

It is easiest to see this in the example of the convex set bounded by the parabola P. The boundary P of this convex set may be illuminated by one direction, that is, $c(F) = 1$ (fig. 81*a*). At the same time, as we see immediately, it is impossible to cover F by any finite number of smaller homothetic sets, that is, $b(F) = \infty$. In fact, let F' be a set homothetic to F with coefficient of homothety $k < 1$, and with centre of homothety O lying outside F (fig. 81*b*). Draw the tangents OA and OB from O to the parabola P bounding

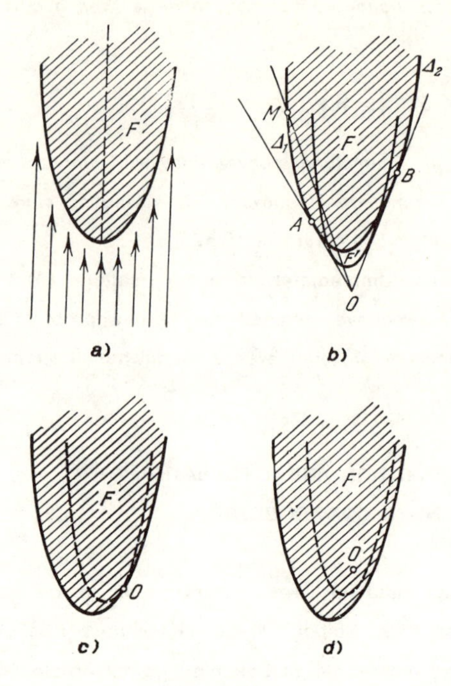

Figure 81.

the set F. The points A and B partition the parabola P into three parts: the arc AB and two infinite arcs Δ_1 and Δ_2, ending at A and B. It is clear that the set F' does not contain any point of the arcs Δ_1 and Δ_2 (because if M is any point of Δ_1 or Δ_2, then there are absolutely no points of F on the line OM beyond M). Thus, the set F' may contain only a finite section of the parabola P (lying on the arc AB). If the centre of homothety O belongs to F, then F' contains no more than one point of the parabola P (fig. 81*c*,*d*). Thus, each set homothetic to F with coefficient $k < 1$ contains only a finite section of the parabola P, and therefore an infinite number of smaller homothetic sets are necessary to cover the whole of F (containing the parabola P), that is, $b(F) = \infty$.

At the same time, there also exist unbounded convex sets for which $b(F)$ is finite. For example, if F is a semi-infinite strip (shaded in fig. 82*a*), then $b(F) = 2$. Notice that in this case, $c(F) = 2$ also, that is, $b(F) = c(F)$.

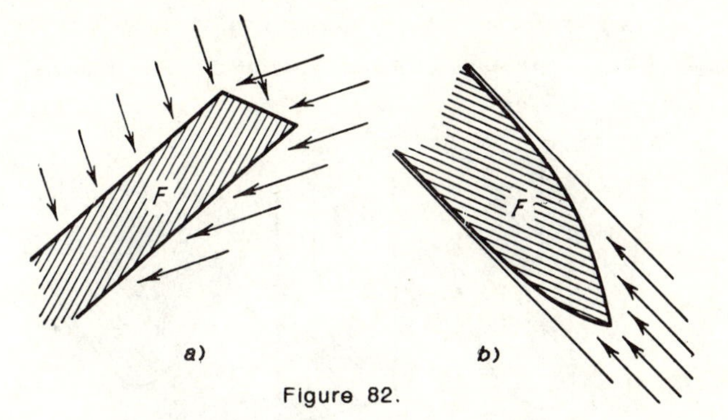

Figure 82.

Lastly, there also exist unbounded convex sets for which both $b(F)$ and $c(F)$ are finite, but differ from each other. For example, if the set F lies in a strip between two parallel lines, and asymptotically tends towards the boundaries of these lines (fig. 82*b*), then, as is easily shown, $b(F) = 2$, $c(F) = 1$.

In connection with the above, the following questions arise:

For what unbounded convex sets does the equality $b(F) = c(F)$ remain true?

For what unbounded convex sets does $b(F)$ take finite values? Do there exist unbounded convex sets for which $c(F) = \infty$? (See Problems 12, 13, 14.)

We answer some of these problems here. First of all we show that from Theorem 7 something nevertheless remains valid also for convex sets. Namely, the first part of the proof of Theorem 7 remains wholly intact, and therefore, *for any unbounded convex set F, we have the inequality:*

$$c(F) \leqslant b(F) \qquad (***)$$

Now we state a theorem proved in 1963 by Soltan (Theorem 10) answering to the second of the questions posed. Let F be an unbounded convex set (of any number of dimensions).

Take an arbitrary interior point of F and consider all possible rays emanating from O which are wholly contained in F. Taken together, all such rays form, as is easily proved, an unbounded region K, called the *inscribed cone* of F with apex at O. For example, for a parabola

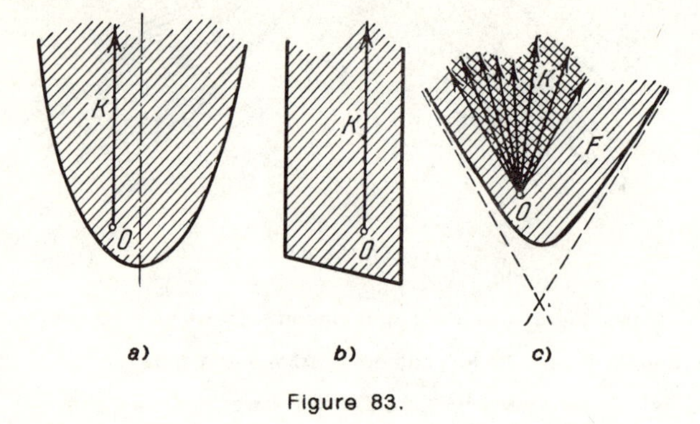

Figure 83.

(fig. 83*a*) or a semi-infinite strip (fig. 83*b*), the inscribed cone consists of only one ray, but for the set shown in Figure 83*c* (the interior region of one branch of a hyperbola), the inscribed cone is represented by an angle. (Notice that if, instead of the point O, we take any other interior point of F as the vertex, the inscribed cone does not change but only undergoes a parallel translation.)

Furthermore, Soltan calls an unbounded convex set F *almost conic* if there exists an r such that all points of F lie at a distance at most r from the inscribed cone K. For example, the sets shown in Figures 83*b* and 83*c* are almost conic. At the same time, the set shown in Figure 83*a* is not almost conic, as points of the parabola recede further and further from its axis.

Theorem 10. *Let F be an unbounded convex set. Then $b(F)$ is finite if and only if F is almost conic.*

Let F be an unbounded n-dimensional convex set which is almost conic, but does not wholly contain any line. We shall denote the dimension of the inscribed conic by q. Soltan constructs a certain bounded $(n-q)$-dimensional convex set M determined by the set F (15), and proves that for this set M,

$$b(F) = b(M)$$

significantly sharpening Theorem 10. In particular, if $q = n$, then M is a point (because $n-q = 0$), and therefore $b(F) = b(M) = 1$, whereas if $q = n - 1$, then M is a line segment (because $n-q = 1$), and therefore $b(F) = b(M) = 2$. Thus, *if an n-dimensional almost conic convex set F (not wholly containing any line) has an n-dimensional inscribed cone, then* $b(F) = 1$, *whereas if it has an (n-1)-dimensional inscribed cone, then* $b(F) = 2$. When applied to plane sets, this gives the following result, found in 1961 by the Soviet mathematician B.N. Visityei [36]. Let F be a two-dimensional almost conic set, not wholly containing any line. If its inscribed cone K is a ray, then $b(F) = 2$, but if K has an apex, $b(F) = 1$. Lastly, if a two-dimensional convex set wholly contains a line, then it can be a strip, a half-plane or a plane. In these cases, $b(F)$ respectively takes the values 2, 1, 1. By the same token, the question about the values of $b(F)$ is fully answered for plane unbounded convex sets.

To conclude this chapter, we note that Soltan constructed examples of three-dimensional unbounded convex sets for which $c(F) = \infty$. The simplest example of this kind is obtained by the following construction. Consider an ordinary circular cone (unbounded) and draw a plane Γ intersecting it, parallel to the generator. This plane cuts the cone into two unbounded convex bodies, of which we shall consider the one containing the cone's apex. This considered unbounded convex body F (fig. 84) has the required property: $c(F) = \infty$.

In fact, let us denote by P the parabola obtained by the intersection of the cone with the plane Γ. Each point A of the parabola P is a corner point of F, and moreover a two-faced corner point of F at A (formed by the plane Γ and the tangential plane of the cone at A) with angle tending to zero as the point A moves off

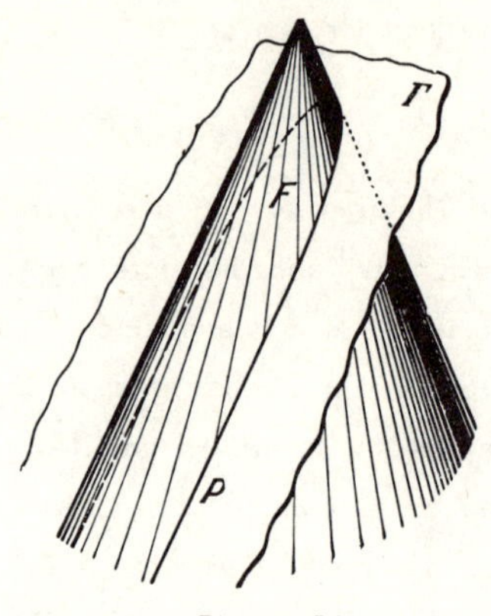

Figure 84.

along the parabola to infinity. From this it is easily deduced that each direction can illuminate only a finite arc of the parabola P lying on the boundary of F. Consequently, to illuminate all the boundary of F (and, in particular, all points of the parabola P), an infinite number of directions are necessary, that is, $c(F) = \infty$.

CHAPTER 3

SOME RELATED PROBLEMS

§17. BORSUK'S PROBLEM FOR NORMED SPACES

If the chosen line segment LM is taken as the unit of length, then the length of an arbitrary segment AB is defined as the ratio $AB:LM$. The length of the segment AB depends only on its magnitude, and certainly does not depend on the direction and position of the segment. However, in certain problems, it is necessary to use a different definition of segment length, in which the length of a segment depends on its magnitude and also on its direction. To define distance in this new sense, we must assign a unit of length in each direction separately. A very interesting definition of this sort was proposed at the end of the 19th century by the well-known German mathematician H. Minkowski. We shall firstly consider this definition restricting ourselves to the case of geometry in the plane.

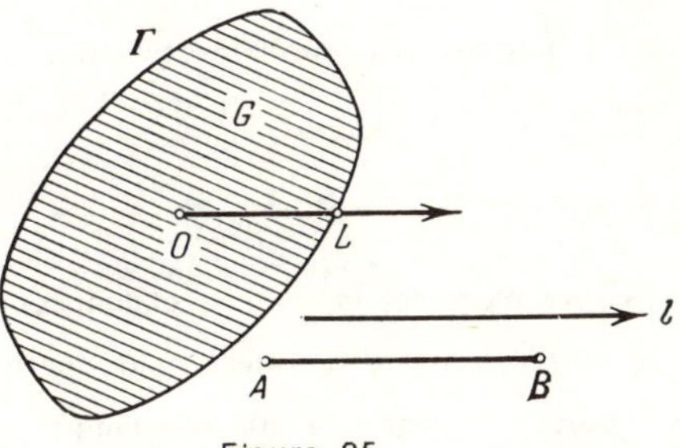

Figure 85.

Suppose we are given a bounded plane convex set G, symmetric with respect to some point O (fig. 85). Denote the curve bounding the set G by Γ. We shall consider the unit of length corresponding to the direction ℓ to be the line segment OL of the ray parallel to ℓ from the point O to the point L of intersection of this ray with the line Γ. The length of the segment AB with respect to the new system of measure is now defined as the ratio $AB:OL$, where OL is the unit of length parallel to the direction ℓ defined by the vector $\overline{AB}$. (In the case when A coincides with B, it is natural to regard the length of the segment AB as being zero.) Later on, we shall denote the length of AB with respect to the new system of measure by $d_G AB$. It is obvious that $d_G OM = 1$ if and only if M lies on the curve Γ. If the point M lies inside the set G, then $d_G OM < 1$, whereas if M lies outside G, $d_G OM > 1$.

Notice that if G is a disc, we arrive at the usual definition of length in which the length of a segment depends only on its magnitude, but not on its direction.

We shall now show the basic properties of the new definition of length. As we already know,

$$d_G AB \geqslant 0$$

Moreover, equality holds if and only if A and B coincide. Furthermore, G being centrally symmetric implies the equality:

$$d_G AB = d_G BA.$$

Lastly, if AB and CD are parallel segments, and also $AB:CD = k$, then:

$$d_G AB = k . d_G CD.$$

Up to now, we have not made use of the convexity of G anywhere. It turns out that the convexity of G guarantees the following very important property of the new length:

<u>The triangle inequality</u>. *In any triangle ABC the length of one of the sides (with respect to the measure defined on G) does not exceed the sums of the lengths of the other two sides.* (In what follows, the triangle inequality will not be useful to us, but for the interested reader we give the proof.)

Proof. Suppose that:

$$d_G BC = a, \qquad d_G AC = b, \qquad d_G AB = c.$$

Further, draw the "radii" OP and OQ in G, having the same directions as the vectors $\overline{BC}$ and $\overline{CA}$ (fig. 86). Next, take the point M on the segment OP such that $OM:MP = a:b$, and in the triangle OPQ draw the segment $MN \parallel OQ$. Taking into account the similarity of the triangles OPQ and MPN, we have:

$$d_G OM = OM:OP = \frac{a}{a+b},$$

$$d_G MN = MN:OQ = MP:OP = \frac{b}{a+b}.$$

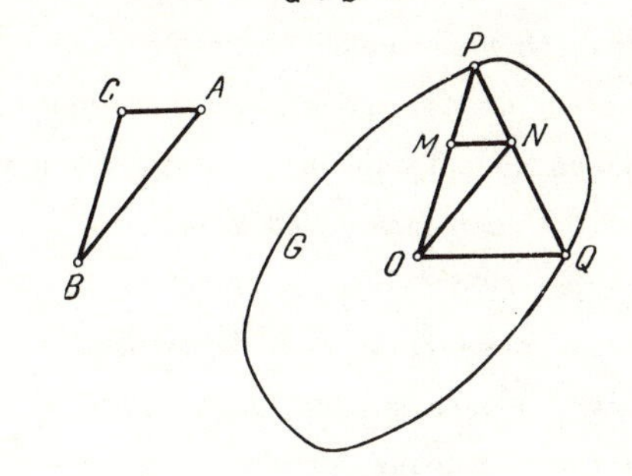

Figure 86

Consequently,

$$BC:OM = d_G BC : d_G OM = a : \frac{a}{a+b} = a + b,$$

$$CA:MN = d_G CA : d_G MN = b : \frac{b}{a+b} = a + b.$$

Thus, $BC:OM = CA:MN$, and in addition, $\angle BCA = \angle OMN$. This implies that the triangles BCA and OMN are similar, and therefore, $AB \parallel ON$ and $AB:ON = a + b$, that is, $d_G AB : d_G ON = a + b$. So:

$$d_G ON = \frac{d_G AB}{a+b} = \frac{c}{a+b}$$

But the points P and Q belong to G. By virtue of the convexity of this set, all the segment PQ belongs to it and, in particular, the point N belongs to G. From this it follows that $d_G ON \leqslant 1$, that is, $\frac{c}{a+b} \leqslant 1$, or, finally, $c \leqslant a + b$. This means that:

$$d_G AB \leqslant d_G BC + d_G AC.$$

A plane in which a measure of length is given by a certain

convex centrally symmetric set is called a *normed plane*. We analogously define *n-dimensional normed space* that is, a space in which the measures of length are given by a convex centrally symmetric body G. This body G is sometimes called the *unit ball* of the normed space.

Let us now turn our attention to Borsuk's problem for normed spaces. The problems considered in the preceding sections suggest substituting the usual plane or space by a normed plane or normed space. As a matter of fact, in the problem of covering by smaller homothetic sets and the illumination problem, the lengths of the segments generally do not appear, and consequently, the statements of these problems do not depend on which way these lengths are defined. Such problems are called *affine*.

The situation with Borsuk's problem is different. In this problem, a set of diameter d must be partitioned into parts of smaller diameter. It is clear that the diameter of the parts, as of the whole set, fundamentally depends on which way the lengths of the segments are defined.

For example, in the usual definition of length, it is possible to partition a parallelogram into two parts of smaller diameter (fig. 12*b*). If this parallelogram is considered in normed space, where it itself plays the role of the set G which gives the measure of length, then the diameter of the whole parallelogram and its indicated parts is obviously equal to two. This is implied by the fact that in the normed plane which we consider, the length of each side and each diagonal of this parallelogram is two. Therefore, in this case it is impossible to partition the parallelogram into three parts of smaller diameter. However, four parts is sufficient for such a partition.

Thus, $a(F)$ depends on the choice of the set G, playing the role of the unit disc in the normed plane. Therefore in what follows, when considering Borsuk's problem for the normed plane (or space), we shall denote $a(F)$ by $a_G(F)$.

The problem of determining the magnitude of $a_G(F)$ was considered in 1957 by B. Grünbaum [16] to whom the fundamental part of Theorem 11 proved below is also due. However, Grünbaum's proof is more complicated than the one stated here.

Theorem 11. *For any plane bounded set F, the following relation holds*:

$$a_G(F) \leqslant 4$$

Moreover, equality is attained only in the case when G and F are homothetic parallelograms.

Proof. First of all, notice that the inequality (*) (§9) remains valid in the case of normed space:

$$a_G(F) \leqslant b(F). \qquad (****)$$

This is established in the same way as inequality (*). Consequently, if the set F is not a parallelogram, by Theorem 5, $a_G(F) \leqslant b(F) = 3$.

Now let F be a parallelogram. Draw two lines parallel to the diagonals of the parallelogram through the centre O of G, and denote the segments of these lines cut off by G by by A_1C_1 and B_2D_2. On the segments A_1C_1 and B_2D_2, as on the diagonals, we construct two parallelograms with sides parallel to the sides of the parallelogram F (fig. 87). Denote the smaller of these two

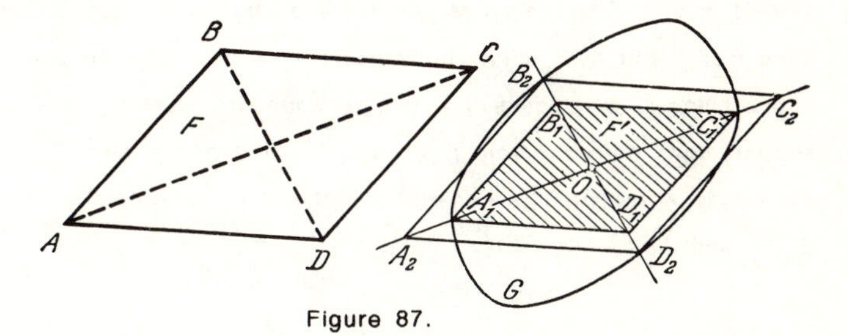

Figure 87.

parallelograms by F', and let this be the parallelogram $A_1B_1C_1D_1$ with diagonal A_1C_1. As the parallelograms F and F' are homothetic, $a_G(F) = a_G(F')$, and therefore in the following argument, we shall be able to consider the parallelogram F' instead of F.

Clearly $d_GA_1C_1 = 2$, and therefore the diameter of F' is less than two. (Here and in what follows, we have in mind "diameters" with respect to the distances defined by G.) But the diameter of F' is not greater than two because it lies wholly within the set G which has diameter 2. So the diameter of F' is equal to two.

Now draw through O the line p which is parallel to the sides A_1B_1 and C_1D_1, and denote the points of its intersection with the

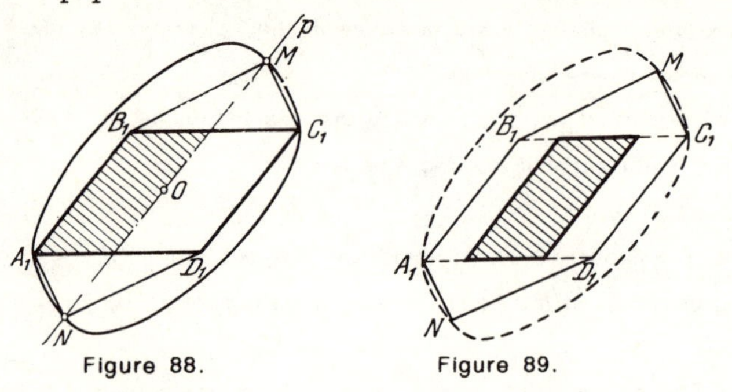

Figure 88. Figure 89.

boundary of G by M and N. If the points M and N do not lie on the sides B_1C_1 and A_1D_1 of the parallelogram F' (fig. 88), then the whole hexagon $A_1B_1MC_1D_1N$ is contained in F. This easily implies that the diameter of each of the halves into which the line p splits the parallelogram F', is less than two (fig. 89), so $a_G(F') = 2$. If the points M and N lie on the sides of F', then the lines B_1C_1 and A_1D_1 must be support lines of G (because a support line G must pass through the boundary point M, and all support lines different from B_1C_1 cut F', and hence also G). Thus, the whole of G is contained in the strip between the line B_1C_1 and A_1D_1 (fig. 90).

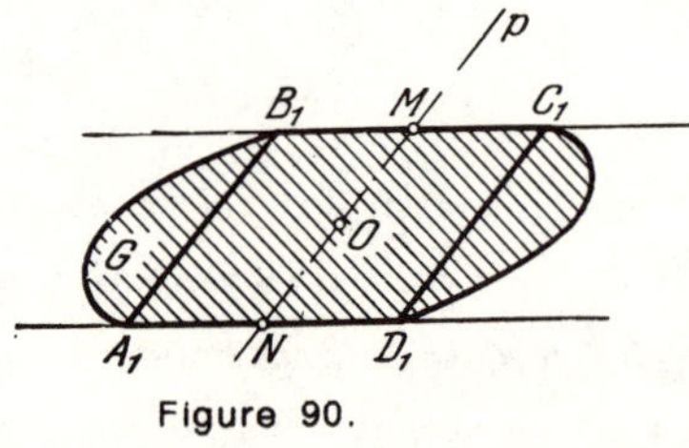

Figure 90.

So either $a_G(F') = 2$, or the set G is contained between the lines B_1C_1 and A_1D_1.

Analogously, drawing the line q parallel to the sides B_1C_1 and A_1D_1, we find that either $a_G(F') = 2$, or G is contained between the lines A_1B_1 and C_1D_1. Combining the two cases, we conclude that either $a_G(F') = 2$, or the set G is contained between both the indicated strips, that is it is surrounded by the parallelogram F' (fig. 91). But in the latter case, G must coincide with the parallelogram F' (because it contains F').

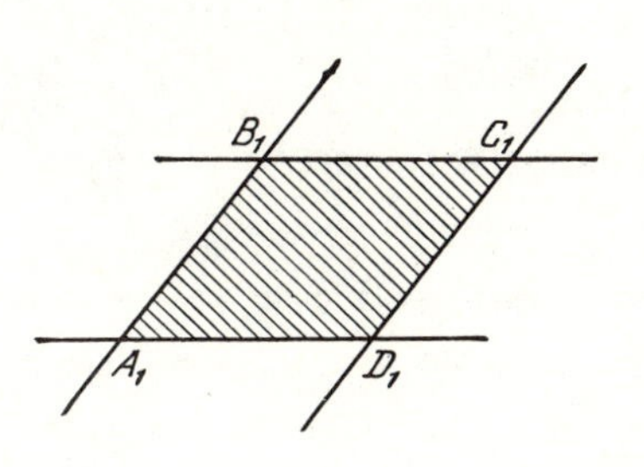

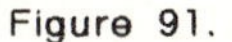
Figure 91.

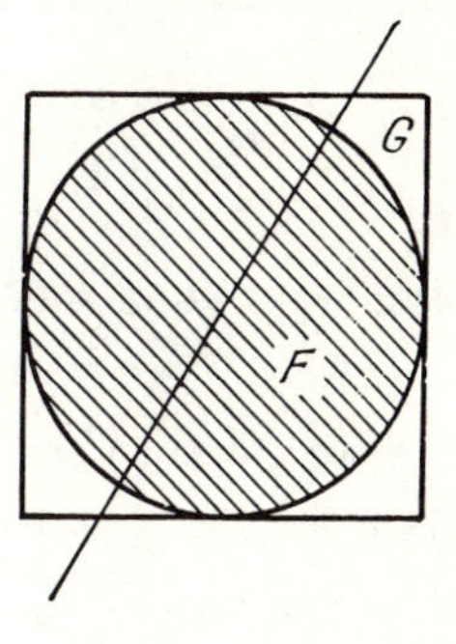

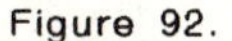
Figure 92.

So either $a_G(F) = a_G(F') = 2$, or the set F coincides with F' (i.e. is homothetic to F) in which case, as we already know, $a_G(F) = a_G(F') = 4$, proving the theorem.

The inequality (****) implies (by virtue of Theorem 9), that *if a body F, in n-space, has a smooth boundary (this condition may be violated in no more than n points), then for any centrally symmetric convex body giving a measure,* $a_G(F) \leqslant n + 1$. However, it is easily seen that if F is an n-dimensional parallelepiped, $a(F) = 2^n$. The inequality (****) and Hadwiger's conjecture reduce to the following conjecture:

For any bounded body F lying in n-dimensional normed space with the unit ball G:

$$a_G(F) \leqslant 2^n$$

with equality holding only in the case when G and F are homothetic parallelepipeds.

Possibly the following relationship holds:

$$a_G(F) \leqslant b(G)$$

(see Problem 15). For the case $n = 2$, this inequality is proved above. But even for $n = 3$ no proof is known. Note also that there are cases for which $a_G(F) < a(F)$. In fact, if G is a square and F is a disc, it is easily seen that $a_G(F) = 2$ (fig. 92), whereas $a(F) = 3$.

§18. THE PROBLEMS OF ERDÖS AND KLEE

Consider an arbitrary n-dimensional rectangular parallelepiped and denote the set of all its vertices by M. Thus, if $n = 2$, M consists of four points (fig. 93); for $n = 3$, M consists of eight points (fig. 94). In general, for an arbitrary natural number n, M

Figure 93.

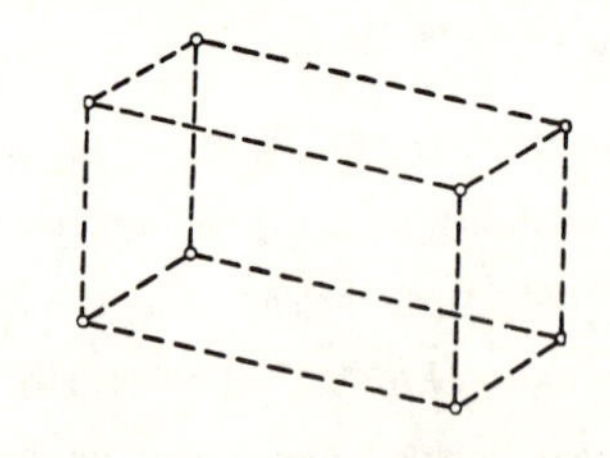

Figure 94.

consists of 2^n points. It is easy to see that if A, B, C are any three points of M (that is, any three vertices of an n-dimensional rectangular parallelepiped), then $\angle ABC$ does not exceed $\pi/2$. In fact, let us take B as the origin of a rectangular coordinate system, and direct the coordinate axes along the edges of the parallelepiped (the case $n = 3$ is shown in fig. 95). Then all the parallelepiped will lie in one corner of the coordinates, and in particular in the one in which all the coordinates are non-negative. Therefore, all the coordinates of the vectors $\underline{BA}$, $\underline{BC}$ (and also their scalar product) are non-negative, and consequently, $\cos \angle ABC \geqslant 0$, that is, $\angle ABC \leqslant \pi/2$.

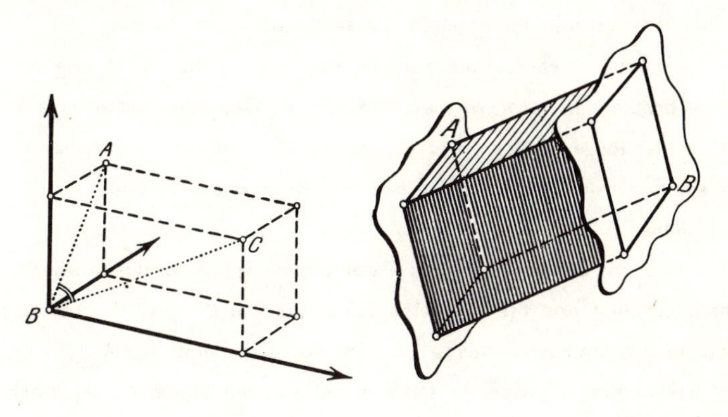

Figure 95. Figure 96.

So, in n-dimensional space there exists a set M consisting of 2^n points (namely, the set of all vertices of a rectangular parallelepiped), such that for an arbitrarily chosen three points A, B, C of this set, $\angle ABC \leqslant \pi/2$. *Does there exist in n-dimensional space a set containing more than 2^n elements with the same property?* This problem was posed (around 1950) by the well-known Hungarian mathematician P. Erdős [11]. He predicted that the answer to this question is negative, that is, that such a set cannot consist of more than 2^n elements.

A problem posed in 1960 in a paper of the American mathematician V. Klee [27] is closely connected to this problem of Erdős, and also arises from considering parallelepipeds. Let us again denote the set of all vertices of a parallelepiped by M (though now, it is not necessarily rectangular). If A and B are two arbitrary points of M, then it is possible to find two opposite faces of the parallelepiped, one of which contains the vertex A, and the other B. Therefore, there exist two parallel support planes of M (the planes of these faces), passing respectively through A and B (fig. 96). In n-space, the situation is analogous, except that it is necessary to consider support hyperplanes.

So in n-dimensional space, there exists a set M consisting of 2^n points such that for any two points A and B, there exist two parallel support hyperplanes passing respectively through A and B. Klee's problem is to *prove that in n-space there does not exist a set having the above property and consisting of more than 2^n elements*; (in addition, we only consider sets not lying wholly in one hyperplane).

It is not difficult to find the connection which exists between Klee's problem and the problems considered in Chapter 2. Let N be a set in n-dimensional space, consisting of m points $A_1, A_2, \ldots, A_m$ and having Klee's property, (that is for any two points A_i, A_j there exist two support hyperplanes passing through them). Denote the convex hull of the points of N by V. Then we have $b(V) = m$.

In fact, let A_i, A_j be any two points of N, and let α, β be two parallel support hyperplanes of N passing through A_i and A_j respectively. Clearly, α and β will also be support hyperplanes of V. Now let V' be a body homothetic to V with coefficient $k < 1$, and let α' and β' be support hyperplanes of V', parallel to α and β. Then the distance between the hyperplanes α' and β' is less than between α and β, and therefore both points A_i, A_j cannot be contained in the strip bounded by the hyperplanes α' and β'. Moreover, the body V' cannot simultaneously contain both points A_i, A_j. From this it is clear that V' contains no more than one of the points $A_1, A_2, \ldots, A_m$. So, to cover V with smaller homothetic bodies, each of the points $A_1, A_2, \ldots, A_m$ must be covered by a separate body, and therefore the total number of smaller homothetic bodies covering V is at least m, that is, $b(V) \geqslant m$. On the other hand, if we choose m directions illuminating the vertices $A_1, A_2, \ldots, A_m$ of the body V, then they will obviously illuminate the whole surface of the polytope V. Consequently, $c(V) \leqslant m$, and therefore $b(V) \leqslant m$. What has been proved implies the equality $b(V) = c(V) = m$.

It can be shown that the connection between Klee's problem and the number $b(V)$ will help in solving Hadwiger's problem. More precisely, this means the following. If it were possible to get a negative answer to Klee's problem, that is, to construct a set N having Klee's property and consisting of more than 2^n points, then for the convex hull V of N, we would have $b(V) > 2^n$, that is, we would get a negative answer to Hadwiger's problem. However, a positive answer to Klee's problem would tell us nothing about the solution of Hadwiger's problem because it would mean only that $b(V) \leqslant 2^n$ for certain specific types of body V (but not for any convex n-dimensional bodies). See Problem 11 in connection with this.

What has happened about the problems of Erdős and Klee? In 1962, it was proved by L. Danzer and B. Grünbaum [6] that both problems have positive solutions, that is, the following holds:

Theorem 12. *A set N, lying in n-space and having the property of Erdős or Klee, contains no more than 2^n elements.*

Proof. Firstly we shall show that if the set N has Erdős' *property, then it also has Klee's property.* Let the set N consisting of the points $A_1, A_2, \ldots, A_m$ have Erdős' property. Taking two arbitrary points A_i, A_j of N, draw two hyperplanes perpendicular to the segment

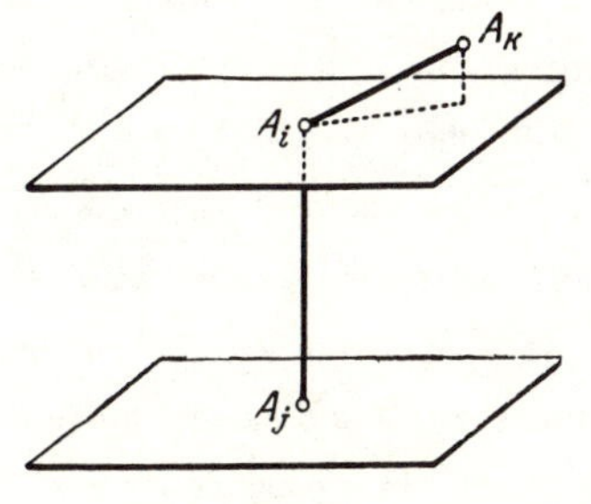

Figure 97.

A_iA_j through these points (fig. 97). Then these hyperplanes must be support planes of N, for if there were a point A_k lying on the opposite side to A_j of the hyperplane passing through A_i, then

$\angle A_jA_iA_k$ would be obtuse, contradicting Erdős' property. So through any two points A_i, A_j of N, there exist two parallel support planes of N, that is, N has Klee's property.

It remains to prove that in n-space, any set having Klee's property contains no more than 2^n points. Let the set N, consisting of the points $A_1, A_2, \ldots, A_m$, have Klee's property. Denote the convex hull of the points of N by V_1. Further, denote by $V_2, \ldots, V_m$ the bodies obtained from V_1 by parallel translations corresponding to the vectors $\overline{A_1A_2}, \ldots, \overline{A_1A_m}$ (fig. 98).

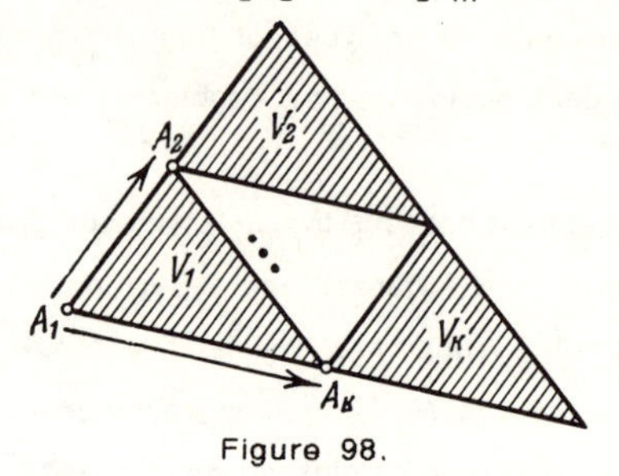

Figure 98.

Firstly, we shall show that the bodies $V_1, V_2, \ldots, V_m$ pairwise have no interior points. In fact, consider the bodies V_i and V_j. Draw mutually parallel support hyperlanes α and β through A_i and A_j (belonging to V_1). For convenience, let us consider these hyperplanes as being horizontal, and regard V_1 as lying below the hyperplane α and above β (fig. 99). Furthermore, choose a point B such that $\overline{A_1A_j} = \overline{A_iB}$, then also $\overline{A_1A_i} = \overline{A_jB}$. In other words, the parallel translation along the vector $\overline{A_1A_j}$ maps A_i to B and the hyperplane α into the hyperplane γ which passes through B and is parallel to α. This maps the body V_1 into the body V_j, which, consequently, lies below γ. Analogously, the translation along the vector $\overline{A_1A_i}$ maps A_j to B, the hyperplane α into the same hyperplane γ, and V_1 into V_i, which therefore lies above γ. So V_i and V_j lie on opposite sides of γ, and therefore do not have common interior points. (This reasoning is also applicable in the case when one of the indices i, j is 1, because the body V_1 is

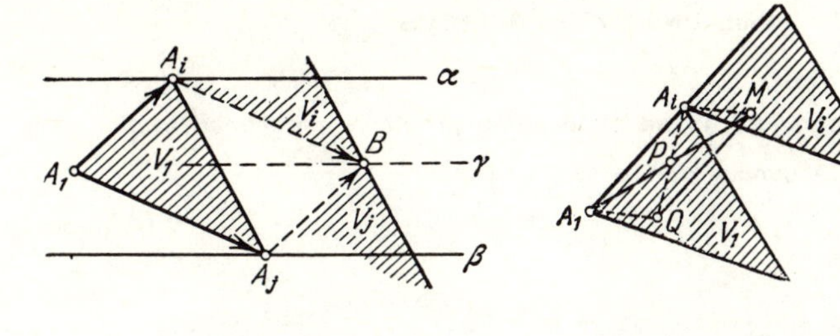

Figure 99. Figure 100.

obtained from V_1 by a translation along the null vector $\overline{A_1A_1}$, just as V_i is obtained from V_1 by a translation along the vector $\overline{A_1A_i}$).

Now let us denote by V the body obtained from V_1 by homothety with centre A_1 and coefficient 2. We shall show that all the bodies $V_1, V_2, \ldots, V_m$ are contained in V. Let M be a point of V_i. Then there exists a point Q of V_1 such that $\overline{QM} = \overline{A_1A_i}$ (fig. 100). Denote the middle of the segment MA_1 by P. Then P also coincides with the middle of the segment QA_i, and therefore, by virtue of the convexity of V_1, P belongs to V_1. As clearly $\overline{A_1M} = 2.\overline{A_1P}$, then from the above, it is clear that M belongs to V, that is, that V_i is contained in V.

Now let v be the volume of the (n-dimensional) body V_1. As the set N does not lie in one hyperplane, $v \neq 0$. Each of the bodies $V_2, \ldots, V_m$ have the same volume v. Furthermore, $V_1, V_2, \ldots, V_m$ taken pairwise do not have common interior points. Therefore, the total volume occupied by $V_1, V_2, \ldots, V_m$ equals mv. Furthermore, the volume of V obtained (in n-space) from V_1 by homothety with coefficient 2, equals $2^n v$. As all the bodies $V_1, V_2, \ldots, V_m$ are contained in V, we have $mv \leqslant 2^n v$. Lastly, recalling that $v \neq 0$, we get the required inequality $m \leqslant 2^n$.

§19. SOME UNSOLVED PROBLEMS

In conclusion, we state some problems mentioned earlier, and others of a similar nature.

Problem 1

We begin with the basic problem of the first chapter, namely Borsuk's problem:

Prove that any body F of diameter d lying in n-space may be partitioned into n + 1 parts of smaller diameter, that is, that in n-dimensional space, $a(F) \leqslant n + 1$.

It is sufficient to solve this problem only for convex bodies, that is, it is sufficient to *prove that any n-dimensional convex body of diameter d may be partitioned into n + 1 parts of smaller diameter*.

We recall that this problem has not been solved for $n \geqslant 4$.

Problem 2

There exists another problem, entirely equivalent to Borsuk's problem and connected with considering the *set of permanent width*.

Let F be any plane set and ℓ be some line. Draw two support lines of F perpendicular to ℓ. The distance h between these support lines is called the *width* of F in the direction ℓ (fig. 101). A convex set F is called a *set of permanent width* if in any direction it has one and the same width. Apart from the disc of diameter d, there exist infinitely many other sets of permanent width, the simplest of which is a *Pello triangle*, bounded by three arcs of a disc of radius d (fig. 102). Analogously, a convex spatial body is

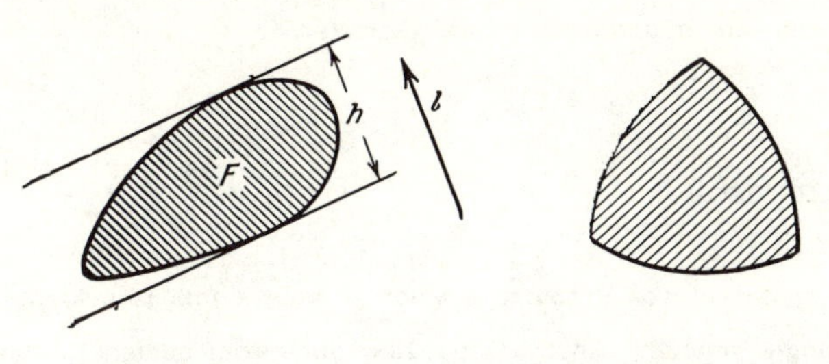

Figure 101. Figure 102.

called a *body of permanent width* if its width (that is, the distance between two parallel support planes), is the same in all directions.

It is easily seen that any set (or body) of permanent width d has diameter equal to d. The converse, of course, is not true, for not every set of diameter d is a set of permanent width d. However, the following important theorem holds (see, for example, the book by Bonnensen and Fenchel [2]):

Any two-dimensional set of diameter d may be covered by some set of permanent width d. Analogously, the theorem holds in three-dimensional space (or in n-dimensional space): *any body of diameter d may be covered by any body of permanent width d.*

It immediately follows from the statement of the theorem that to prove Borsuk's conjecture, it is sufficient to establish the conjecture's validity only for bodies of permanent width. In other words, we proceed to the following problem, equivalent to Borsuk's problem.

> *Prove that any n-dimensional body of permanent width d may be partitioned into $n + 1$ parts of diameter less than d.*

Connected with the mention of bodies of permanent width, we note that in 1955, Lenz [28] proved the following theorem:

No n-dimensional body of permanent width may be partitioned into n parts of smaller diameter. If an n-dimensional convex body with smooth boundary is not a body of permanent width, it may be *partitioned into n parts of smaller diameter.*

Problem 3

Problem 1 (or Problem 2 which is equivalent) is evidently exceedingly difficult. Another (possibly somewhat easier) problem is interesting in this regard:

Prove that any n-dimensional convex polytope of diameter d may be partitioned into n+1 parts of smaller diameter.

The problem has not been solved for $n \geqslant 4$. Note also that when considering the diameter of a polytope, it is sufficient to take only the vertices into consideration. Therefore, the statement of the problem is equivalent to the following:

A finite set of points having diameter d is given in n-space. Prove that this set may be partitioned into n+1 subsets, each of which has diameter less than d.

This problem is interesting in itself, independent of Borsuk's general problem. A simple solution of this problem for $n = 2$ is implied by one of the theorems proved by Erdős [10] in 1946 (see [24]). For $n = 3$, the Hungarian mathematicians A. Heppes and P. Révész found a solution of this problem in 1956 which is significantly simpler than the general solution of Borsuk's problem for $n = 3$, given by Eggleston and Grünbaum. However, this solution was found after the publication of Eggleston's paper, mentioned in §5. For $n \geqslant 4$, the problem has not been solved.

Problem 4

From the proof of Theorem 1 (§3), it is easily deduced that *any plane set of diameter d may be partitioned into three parts, the diameter of each of which does not exceed $d\sqrt{3}/2 \approx 0.8660d$* (as the equality $PL = d$ easily implies that $PQ = d\sqrt{3}/2$; see fig. 18). This bound on the diameter of the parts is best possible [13], because as is easily seen, *it is impossible to partition the disc into three parts each having diameter less than $d\sqrt{3}/2$.*

(In fact, a part having diameter less than $d\sqrt{3}/2$ cuts off a closed set on the circumference (1), the extreme points of which are separated from each other by a angular distance less than 120°; therefore three such parts do not cover the whole circumference.)

At the same time, we noticed in §5 that the bound

$$\frac{\sqrt{6129030-937419\sqrt{3}}}{1518\sqrt{2}}\, d \approx 0.9887d.$$

is not the best possible. Our next problem is to *find the best bound for the diameter of the parts in the problem about the partition of a three-dimensional body of diameter d into parts of smaller diameter.* In other words, the question is about determining the number $\alpha < 1$ such that any three-dimensional body of diameter d may be partitioned into four parts having diameters at most αd, but there exists a three-dimensional body of diameter d which cannot be partitioned into four parts having diameter less than αd. The American mathematician D. Gale [13] conjectured in 1953 that:

$$\alpha = \frac{\sqrt{3+\sqrt{3}}}{\sqrt{6}} \approx 0.888.$$

Up to now, this conjecture has been neither proved nor refuted.

Problem 5

This problem is concerned with the generalization of Borsuk's problem in several directions.*

Let us call the *radius* of an n-dimensional body the radius of its circumscribing n-dimensional ball (that is, the smallest ball containing the given body). It is clear that given an arbitrary body, the radius, generally speaking, will not equal half the diameter. For example, in the case of an equilateral triangle of radius r and diameter d, $d = r\sqrt{3}$.

Our following problem is to *prove that any n-dimensional body of diameter d may be partitioned into n + 1 parts, each of which has radius* less than $d/2$. It is clear that this is a strengthening of Borsuk's problem (because any body of radius less than $d/2$ clearly has diameter less than d). If we denote by $a'(F)$ the smallest number of parts of radius less than $d/2$ into which it is possible to partition the given body F of diameter d, then the problem may be stated as follows:

Prove that for any n-dimensional body F, $a'(F) \leqslant n + 1$.

The fact that the stated problem is a strengthening of Borsuk's problem is obviously shown by the inequality $a(F) \leqslant a'(F)$.

Notice that for $n = 2$, the solution of this problem also gives Theorem 1. It is easily seen that the radius of the three parts into which the regular hexagon in Figure 18 is partitioned is $d\sqrt{3}/4$, that is, less than $d/2$. (Grünbaum's partition proposed in Theorem 3 does not imply that the radius of the parts is less than $d/2$.) Notice also that for an n-dimensional ball of diameter d, a partition into n+1 parts of radius less than $d/2$ is possible, it being sufficient

----

*The reader may find other variants of Borsuk's problem and other unsolved problems in the interesting paper by Grünbaum [18].

to take the partition analogous to that shown in fig. 28.

Problem 6

Let F and G be two n-dimensional bodies. We shall say that G is of *smaller width* than F if the width of G is less than the width of F in any direction. Let us denote the smallest number of parts of smaller width into which it is possible to partition F by $b'(F)$. It can readily be proved that $a(F) \leqslant b'(F) \leqslant b(F)$. Our next problem now consists in *proving that any n-dimensional body F may be partitioned into 2^n parts of smaller width, that is, $b(F) \leqslant 2^n$.*

Notice that this problem is affine, that is, $b'(F)$ does not change with an affine representation of F. As we know, $a(F)$ and $a'(F)$ do not have this property.

Problem 7

For what class of convex bodies F is it true that:

$$a(F) = b'(F) = b(F) \ ?$$

In particular, are these equalities true for bodies of permanent width?

It should not be supposed that $b(F)$ and $b'(F)$ are trivially equal, that is, that a part of F has smaller width than F if and only if it has smaller size. For example, if F is a disc, and its part G is an inscribed equilateral triangle (fig. 53), then G has a smaller width, but the size of G is 1.

Problem 8

The essence of the following problem reminds us of Hadwiger's conjecture.

Prove that any n-dimensional bounded convex body F may be covered by 2^n smaller homothetic bodies (or, equivalently, may be partitioned into 2^n parts of smaller size), that is, $b(F) \leqslant 2^n$. *Prove also that equality is attained only for the n-dimensional parallelepiped.*

Previously (Theorem 5), we had the solutions of these problems for $n = 2$. The solution is not known even for $n = 3$. Furthermore, the solution of this problem is not known even for n-dimensional polytopes:

Prove that any convex three-dimensional polytope N may be covered by eight smaller homothetic polytopes (and, if M is not a parallelepiped, by seven polytopes).

Problem 9

The equivalence theorem (Theorem 7) allows us to state problem 8 differently, equivalent to the above:

Prove that the boundary of any n-dimensional convex body F may be illuminated by 2^n directions, that is, $c(F) \leqslant 2^n$. Prove also that the equality $c(F) = 2^n$ is attained only for the n-dimensional parallelepiped.

This problem remains open even for $n = 3$ and even for three-dimensional polytopes:

Prove that the boundary of any convex three-dimensional polytope M may be illuminated by n directions (and if M is not a parallelepiped, by seven directions).

Problem 10

As we know (see the note on page 62), *if an n-dimensional convex body has no more than n corner points, then* $c(F) = n + 1$. In connection with this, the following problem arises:

Is it true that if F is an n-dimensional convex body F having $n + 1$ *corners,* then $c(F) = n + 1$? (For $n = 2$ the solution is given by Theorem 6).

Is it true that if F is an n-dimensional convex body having at most $n - 3$ corner points, *has cardinality* at most $n - 3$, $c(F) = n + 1$? *In particular, is it true that for any three-dimensional convex body F having only a finite number of corner points* $c(F) = 4$?

Problem 11

Let F be an arbitrary bounded n-dimensional convex body.

Prove that it is possible to choose $c(F)$ *points* $A_1, A_2, \ldots, A_{c(F)}$ *on the boundary of F such that no two of these points may be illuminated by the same direction.*

Notice that if we were to succeed in proving this, it would give a solution to Problem 9. In fact, let us denote by *M* the polytope with vertices $A_1, A_2, \ldots, A_{c(F)}$. As *M* is contained in *F* then, a fortiori, no two vertices of the polytope *M* may be illuminated by the same direction. Therefore, if we "cut off" the convex polyhedral corners of M at the vertices A_i and A_j, and identify the vertices of these polyhedral corners by a translation, then we find that after the translation, the obtained polyhedral corners U_i and U_j have no common interior points (otherwise there would exist a direction illuminating both vertices A_i, A_j). Therefore, the polyhedral corners U_i, U_j may be partitioned by some hyperplane Γ. Drawing hyperplanes parallel to Γ through A_i and A_j, we find that these hyperplanes are supports for M. Thus, the set of points $A_1, A_2 \ldots A_{c(F)}$ has Klee's property (see §18), and therefore, by virtue of Theorem 12, $c(F) \leqslant 2^n$.

Problem 12

The following questions arise in connection with Theorem 10 (§16):

What conditions must an unbounded n-dimensional convex body F satisfy in order that c(F) be finite?

Under what conditions does the equality $b(F) = c(F)$ hold for an unbounded n-dimensional body F?

For plane sets, the solution presents no difficulty. In this case, $c(F)$ always takes a finite value, namely 2 if the boundary of F contains two parallel rays, and 1 in all the remaining cases (if F is not the whole plane, in which case the illumination problem has no meaning).

For $n \geqslant 3$, these problems have not been solved.

Problem 13

Denote by $c'(F)$ the smallest number of point sources of light lying in n-space outside F with which it is possible to illuminate the whole boundary of F (fig. 103). As Soltan has shown, $c'(F)$

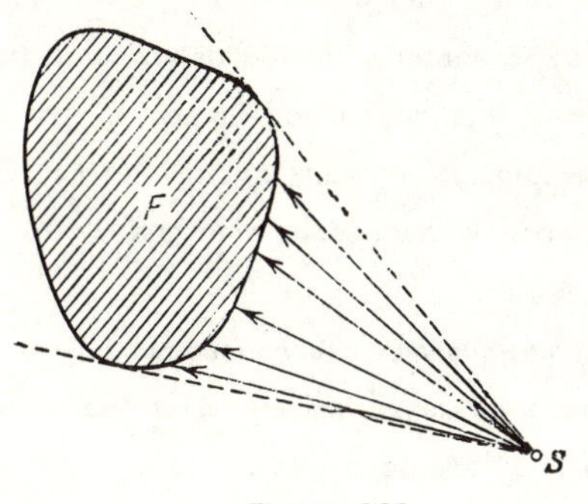

Figure 103.

satisfies $c(F) \leqslant c'(F) \leqslant b(F)$ so that for bounded convex bodies, all three values $c(F)$, $c'(F)$, $b(F)$ coincide. For unbounded bodies, generally speaking $c'(F)$ does not coincide with either $b(F)$ or $c(F)$. The following questions arise in connection with this:

> *What conditions must an unbounded convex set F satisfy in order that $c'(F)$ is finite? Under what conditions do the equalities $c'(F) = c(F)$, $c'(F) = b(F)$ hold?*

For n = 2, this problem was solved by Visityei [36] who proved that $c'(F) = b(F)$ for any plane unbounded set.

Problem 14

Denote by $b''(F)$ the minimum number of convex bodies obtained from the given n-dimensional convex body F by parallel translations, and having the property that their interiors cover the whole of F. The problem of determining $b''(F)$ was posed in 1954 by the German mathematician F. Levi [29] who proved [30] that *if a plane bounded convex set F is not a parallelogram, then* $b''(F) = 3$, *and for a parallelogram,* $b''(F) = 4$. As Soltan [35] proved, $b''(F)$ satisfies the inequality $c(F) \leqslant b''(F) \leqslant b(F)$, so that for bounded convex bodies, all of $b(F)$, $c(F)$, $b''(F)$, $c'(F)$ coincide. Therefore, Theorem 5 immediately follows from Levi's theorem stated above. For unbounded bodies, $b''(F)$, generally speaking, does not coincide with any of $b(F)$, $c(F)$, $c'(F)$. The following questions arise in connection with this:

What conditions must an unbounded convex n-dimensional body satisfy in order to ensure that $b''(F)$ *is finite? Under what conditions do the equalities* $b''(F) = b(F)$, $b''(F) = c(F)$, $b''(F) = c'(F)$ *hold?*

Problem 15

Let us recall some problems connected with normed geometry (pages 69–76):

Prove that for any bounded body F in an n-dimensional normed space with unit ball G:

$$a_G(F) \leqslant 2^n$$

moreover, the equality $a_G(F) = 2^n$ *holds only in the case when G and F are homothetic parallelepipeds.*

Does the inequality $a_G(F) \leqslant b(G)$ *hold for any body F?*

The reader may like to consider Problems 2–5 for normed spaces.

Problem 16

Let F be a bounded convex n-dimensional body. Denote by $i(F)$ the maximum integer having the following property: there exist bodies $F_1, F_2, \ldots, F_{i(F)}$, obtained from F by parallel translations, such that $F_1, F_2, \ldots, F_{i(F)}$ do not overlap, but each of them shares

Figure 104.

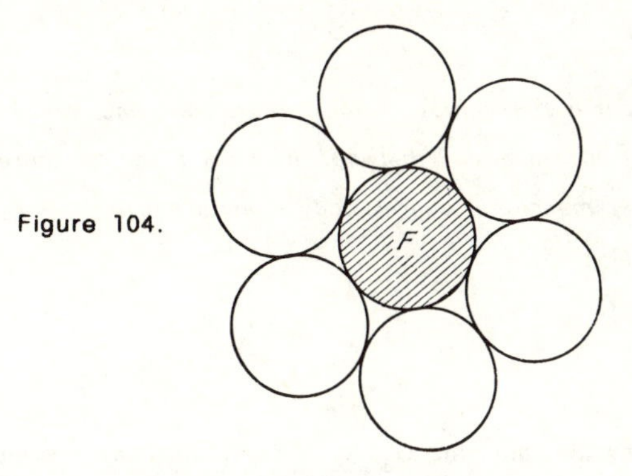

at least one common point with F. For example, if F is a disc, then $i(F) = 7$ (fig. 104), whereas if F is a parallelogram, $i(F) = 9$ (fig. 105). In both cases, one of the sets $F_1, F_2, \ldots, F_{i(F)}$ coincides with the original set F (shaded in the diagram), and the remaining ones "surround" it. In 1961, Grünbaum proved [17] that

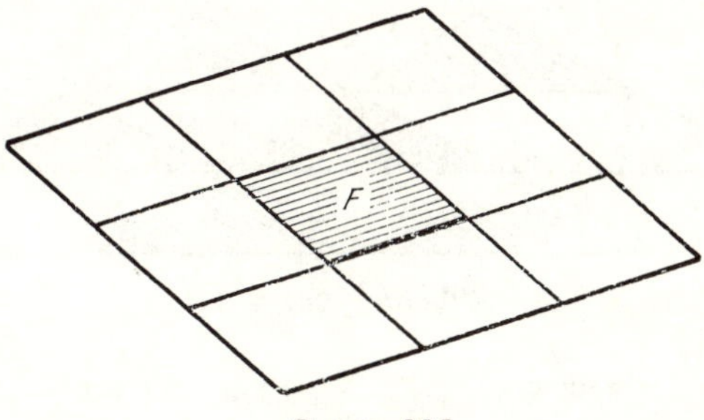

Figure 105.

if a plane convex set F is not a parallelogram, then $i(F) = 7$, and moreover, one of the sets $F_1, F_2, \ldots, F_{i(F)}$ necessarily coincides with the original set F. Furthermore, the following well-known inequalities hold for any n-dimensional convex body F:

$i(F) \geqslant n^2 + n + 1$, equality being attained, for example, for an n-dimensional simplex (Grünbaum [17]).

$i(F) \leqslant 3^n$, equality being attained only for the n-dimensional parallelepiped (Hadwiger [22]).

In connection with this, the following problem was posed by Grünbaum:

Prove that $i(F)$ always takes odd values, and moreover that for any odd number k between $n^2 + n + 1$ and 3^n there exists an n-dimensional convex body F satisfying $i(F) = k$.

Problem 17

We shall say that the sets $G_1, G_2, \ldots, G_k$ *surround* the set F if each line starting from any point of F and extending to ∞ necessarily has a common point with at least one of the sets $G_1, G_2, \ldots, G_k$. Furthermore, we shall denote by $e(F)$ the smallest integer having the following property: there exist bodies $F_1, F_2, \ldots, F_{e(F)}$ obtained from F by parallel translations such that none of $F_1, F_2, \ldots, F_{e(F)}$

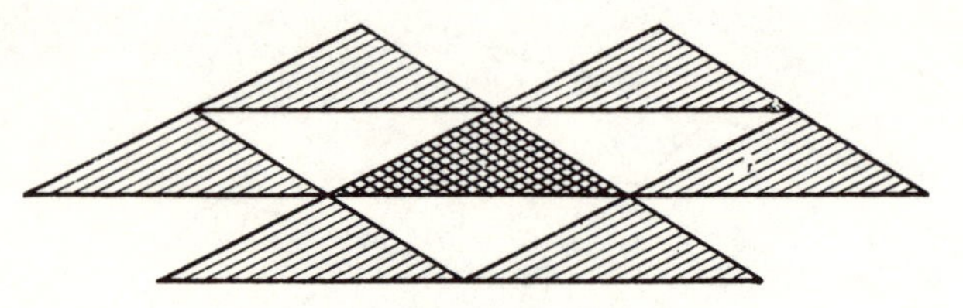

Figure 106.

overlap F and, in addition, $F_1, F_2, \ldots, F_{e(F)}$ surround F. Grünbaum proved in 1961 [17] that *if a plane set F is not a parallelogram, then $e(F) = 6$* (figs. 104 and 106), *and for the parallelogram, $e(F) = 4$* (fig. 107).

What values can $e(F)$ take for an n-dimensional convex

body F?

No results pointing towards a solution of this problem for $n \geqslant 3$ are known.

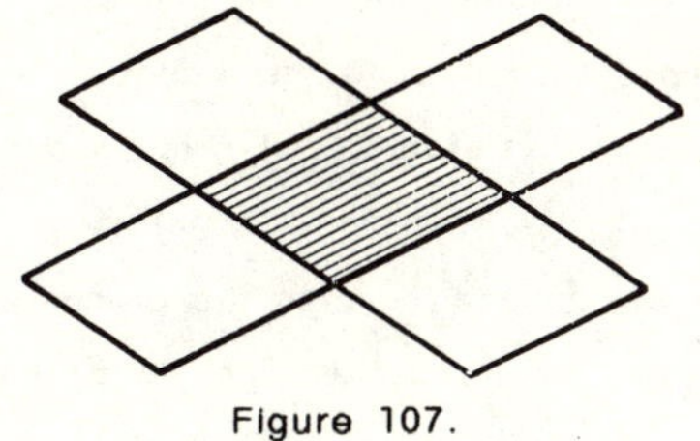

Figure 107.

NOTES

(1) (from page 1). The definition of the diameter of a set mentioned in the text implicitly supposes that each set considered represents a closed set (that is, all its boundary points are included as points of the set). For example, if F is an open disc of diameter d (that is, a disc in which the points round its circumference are not included), then the precise upper bound for the distances between two points of F equals d; however, in this case, there do not exist two points of F at distance exactly d. If we add all the boundary points to the set F (i.e. consider a closed disc), then the exact upper bound will be attained; two points A and B can be found at distance exactly d.

In general, if F is a closed bounded set (in the plane or in Euclidean space of an arbitrary number of dimensions), then two points A and B of F can be found at maximum distance. Actually, let M and N be two arbitrary points of F, and $\rho(M,N)$ be the distance between them. The function $\rho(M,N)$ is continuous (in M and N). But any continuous function (in this case of two variables M and N) with arguments varying in a closed bounded set, must attain its greatest (and smallest) values. Thus, two points A and B of F can be found such that $\rho(A,B) \geqslant \rho(M,N)$ for any points M, N of the set F. The distance $d = \rho(A,B)$ between two such points represents the *diameter* of the set F.

(2) (from page 3). Here, the question is about the partition of a set into parts, and about the diameters of these parts. Corresponding to the previous note, we shall reckon that the parts into which the set is partitioned are themselves closed sets. Therefore, the assertion explained in the text is more precise in the following form: *if a disc F of diameter d is somehow represented as the union of two of its closed subsets, then at least one of these subsets has the same diameter d.* The reasoning mentioned on page 3 does not, of course, fully prove this assertion. The correct proof is as follows. Let us denote by H_1 and H_2 the closed subsets

which we consider, (so that their union $H_1 \cup H_2$ gives the whole disc F). The points of H_1 lying on the circumference of F, make up some set K_1; the set K_2 is defined analogously. Therefore, the circumference of the disc F is represented by the union of two of its closed subsets K_1 and K_2. If one of these sets, for example K_2, is empty (that is, contains no points at all), then K_1 coincides with the whole of the circumference; therefore, the set K_1, and hence also H_1, has diameter d. If both the sets K_1, and K_2 are non-empty, then they must have a common point A (because the circumference is connected and therefore cannot be represented as the union of two non-intersecting convex subsets). Denote by B the point diametrically opposite A, and without loss of generality, let B belong to K_2. Then K_2 contains both points A and B. Consequently, the set K_2, and hence also H_2, has diameter d. And so, in either case, at least one of the sets H_1, H_2 has diameter d.

(3) (from page 4). We shall make one more note as regards the partition of a set into parts. It is possible to understand the word "partition" in the sense that the set F is represented as the union of several of its closed subsets: $F = H_1 \cup H_2 \cup \ldots \cup H_m$ (namely, as in note (2)). In this case, mathematicians usually say that the sets $H_1, H_2, \ldots, H_m$ form a *cover* of the set F. However, it is more natural to understand the term "partition" in the sense of the closed sets $H_1, H_2, \ldots, H_m$ not only making up a cover of F, but furthermore not intersecting each other, that is, pairwise having no common interior points.

It is easy to see that the meaning of the problem about the partition of a set into parts of smaller diameter does not change depending on which of these two meanings we give to the term "partition into parts". In fact, if the set F is represented as the union of several of its closed subsets:

$$F = H_1 \cup H_2 \cup \ldots \cup H_m$$

(possibly overlapping one another), then we can "trim" these parts, without decreasing the diameter, so that they do not overlap each other. For this, notice that the sets*

$$H_1 \backslash (H_2 \cup H_3 \cup \ldots \cup H_m),$$
$$H_2 \backslash (H_3 \cup \ldots \cup H_m),$$
$$\ldots\ldots\ldots$$
$$H_{m-2} \backslash (H_{m-1} \cup H_m),$$
$$H_{m-1} \backslash H_m,$$
$$H_m$$

make up a cover of F, and pairwise have no common points. It is true that these sets may turn out not to be closed. However, the closure of these sets, that is, the sets:

$$H'_1 = \overline{H_1 \backslash (H_2 \cup H_3 \cup \ldots \cup H_m)},$$
$$H'_2 = \overline{H_2 \backslash (H_3 \cup \ldots \cup H_m)},$$
$$\ldots\ldots\ldots$$
$$H'_{m-1} = \overline{H_{m-1} \backslash H_m},$$
$$H'_m = H_m$$

are closed subsets of F pairwise having no common interior points, and giving a cover of F.

Thus, from the arbitrary cover $(H_1, H_2, \ldots, H_m)$ of F by closed subsets, we obtain a cover $(H'_1, H'_2, \ldots, H'_m)$ consisting of sets not overlapping one another. In addition, the diameters of these parts have, of course, not increased (because the set H'_i is contained in H_i).

(4) (from page 5). The reasoning in the text (concerned with the "approaching" of the line ℓ towards the figure F) does not constitute, of course, a rigorous proof of the existence of the support line ℓ_1. It is possible to get a rigorous proof, for example,

---

*The symbol $A \backslash B$ denotes the set obtained by the removal from A of all points belonging to B.

as follows. Draw the line ℓ not intersecting F, and a line $m \perp \ell$. Let us assume that the lines ℓ and m are along the coordinate axes

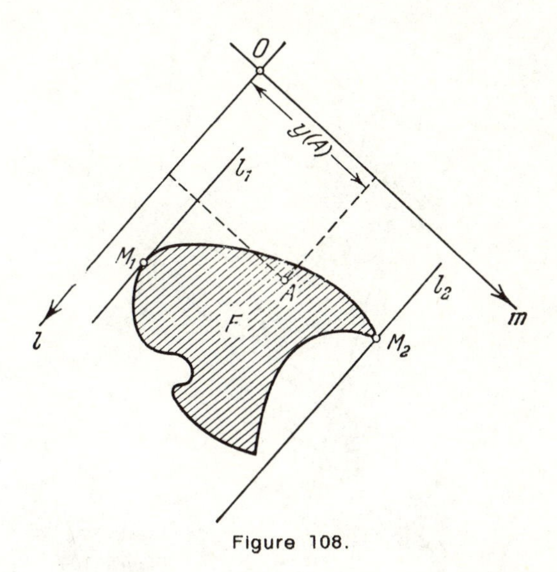

Figure 108.

(fig. 108), and for any point A of F, denote its ordinate (measured along the line m) by $y(A)$. Thus, a function $y(A)$ is defined on the set F, and moreover, this function is continuous (because the difference $y(A)-y(A')$ does not exceed the length of the segment AA'). But a continuous function defined on a closed bounded set F attains its maximum and minimum values. In other words, there exist points M_1 and M_2 of F, such that $y(M_1) \leqslant y(A) \leqslant y(M_2)$ for any point A of F. But this means that if we draw through M_1 and M_2 the lines parallel to the abscissa ℓ, then all the set F will be contained in the strip between these lines. Thus, the lines ℓ_1 and ℓ_2, passing through M_1, M_2 parallel to ℓ are two support lines of F, as discussed in the text (see fig. 14).

(5) (from page 6). We shall show as an example that the point A continuously depends on the direction of the line ℓ_1. Let us assume that ℓ_1 has turned through some angle α (fig. 109). At this point, we denote the position of the lines ℓ_1, ℓ_2, m_1, m_2 by ℓ_1', ℓ_2', m_1', m_2' Draw the lines m_A and m_B through A and B

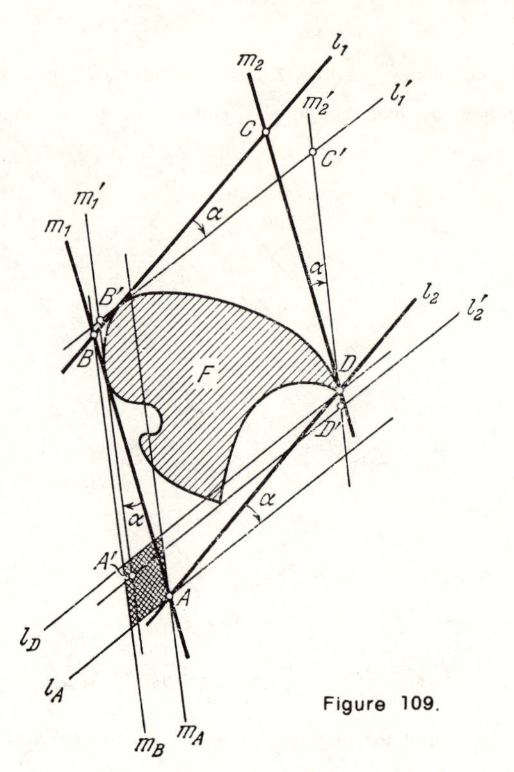

Figure 109.

subtending an angle α with m_1 (that is parallel to m_1'). As the line m_A cuts the set F, whereas m_B does not have any common points with F, the support line m_1' lies between m_A and m_B. Analogously, if we draw the lines ℓ_A and ℓ_D through the points A and D subtending an angle α with ℓ_2 (that is parallel to the line ℓ_2'), then we find that the support line ℓ_2' lies between ℓ_A and ℓ_D. Consequently, the point A' at which the support lines m_1' and ℓ_2' intersect lies inside the parallelogram formed by the lines m_A, m_B, ℓ_A, ℓ_D. But the size of this parallelogram (cross-hatched in fig. 109) may be made arbitrarily small if the angle α is sufficiently small. Thus, the point A' lies arbitrarily close to A if α is sufficiently small. This means that the point A continuously depends on the direction of the line ℓ_1.

An analogous argument shows that the points C, M, N depend continuously on the direction of ℓ_1.

(6) (from page 13). The statement that q closed lines on the ball intersecting neither themselves nor each other partition the ball into $q+1$ parts, is understood well enough. However, a rigorous proof of this assertion is very awkward (it is developed by means of topology). Suffice to say that even for $q = 1$ we obtain the theorem that one simple closed line partitions the ball into two parts; this is the famous Jordan's theorem, the proof of which demands much effort. For us at present, the necessary assertion may be obtained as a trivial consequence of the law of duality by L. C. Pontryagin which, however, we cannot go into in this non-specialized book. Therefore, we shall confine ourselves to the intuitive "understanding" of the assertion.

(7) (from page 15). Let us introduce on the line Γ a parameter t, varying from 0 to 1 as the point runs along the arc Γ from C to C'. The set M of all values of t for which the corresponding point of the arc Γ belongs to the set N_2, is closed (because N_2 is closed). Consequently, M contains a maximum element t_0. In other words, the point of the arc Γ corresponding to the value of the parameter t_0, belongs to the set N_2, and the points of Γ corresponding to the maximum values of the parameter do not belong to N_2. But this means that there exists a last point D of the set N_2, meeting us as we move along Γ from C to C' (namely, the point corresponding to t_0).

Furthermore, if the point D did not belong to the set N_3, then the distance from D to the set N_3 would be positive (recall that N_3 is closed). Therefore, the points of the ball S close to D also would not belong to N_3.

(8) (from page 32). Notice that the sets $N_0, N_1, \ldots, N_n$ are closed. In fact, putting a boundary point $f(A)$ of the ball E in correspondence with a boundary point A of F such that the tangential hyperplanes at these points are parallel, and moreover, the bodies F and E lie on one side of these hyperplanes, we obtain a mapping f of the boundary of F into the boundary of E. This mapping is one-valued and well-defined (because a unique support hyperplane passes through each boundary point of F) and, as can be proved,

is continuous. Further, by definition, the point A belongs to the set N_i if and only if $f(A)$ belongs to M_i, that is, $N_i = f^{-1}(M_i)$. As the mapping f is continuous, and the set M_i is closed, then $N_i = f^{-1}(M_i)$ is also closed $(i = 0, 1, \ldots, n)$.

(9) (from page 41). Let F be a convex bounded set and G be some part of it (that is, a closed subset). Consider all possible sets, homothetic to F with coefficient of homothety at most 1, and containing G. We shall denote by k_0 the greatest lower bound of all the coefficients of homothety for such homothetic sets. If $k_0 = 1$, then the size of the part G equals 1 (because there does not exist a set homothetic to F with coefficient of homothety less than 1 and containing G). Let $k_0 < 1$. Then we may choose a sequence $F_1, F_2, \ldots, F_q, \ldots$ of sets homothetic to F with centres of homothety respectively $k_1, k_2, \ldots, k_q, \ldots$, such that each of these sets contains G and the equality $\lim_{q \to \infty} k_q = k_0$ holds. In addition, we may suppose that we have the inequalities:

$$1 > k_1 > k_2 > \ldots > k_q > \ldots > k_0.$$

It is easy to see that all the points $O_1, O_2, \ldots, O_q, \ldots$ lie at a distance no greater than $d/(1-k_1)$ from the set F (where d is the diameter of F). In fact, let us suppose that the point O_q lies at distance greater than $d/(1-k_1)$ from F. By the homothety with centre O_q and coefficient k_q, the point A is mapped to the point A', such that $O_qA' = k_q.O_qA$. Therefore, we have:

$$AA' = (1-k_q).O_qA > (1-k_1).O_qA > d.$$

Thus, each point A of F is shifted a distance greater than d by the homothety, that is, it is mapped to A' which does not belong to F. In other words, the set F_q, into which the considered homothety maps F, does not have common points with F. But this contradicts the fact that F_q contains the part G of F.

So all the points $O_1, O_2, \ldots O_q, \ldots$ lie at bounded distance from F. Therefore, the sequence $O_1, O_2, \ldots O_q, \ldots$ has at least one limit point. Without loss of generality, we may consider (passing, if necessary, to a subsequence), that the sequence $O_1, O_2, \ldots O_q, \ldots$ has only one limit point O_0, that is, there exists a limit $\lim_{q\to\infty} O_q = O_0$.

It is easy to see that the set F_0, homothetic to F with centre of homothety O_0 and coefficient k_0, contains G (because $\lim_{q\to\infty} k_q = k_0$, $\lim_{q\to\infty} O_q = O_0$). Thus, there exists a set F_0, homothetic to F with coefficient k_0 containing G, but no set homothetic to F with coefficient $< k_0$ can wholly contain G (by definition of the greatest lower bound). This means that k_0 is the size of G. By the same token, it is established that the concept of size is defined for any part G of F.

(10) (from page 57). In the case of an n-dimensional convex body F (for $n > 2$) the *region of illumination* (that is, the set of all points which are points of illumination relative to the direction ℓ) is, of course, not an arc. However, the region of illumination is (for any n) an open subset of the boundary of the body F. In fact, if A is a point of illumination relative to the direction ℓ, then all the boundary points of F close to A will also be points of illumination relative to this direction. But this means that the region of illumination is an open subset of the boundary of F.

(11) (from page 57). In the n-dimensional case, instead of the "reduced arc" described in the text, we shall need to use the following proposition.

Let $\ell'_1, \ell'_2, \ldots, \ell'_s$ be directions sufficient for the illumination of all the boundary of the n-dimensional body F. Denote the region of illumination relative to these directions by $\Delta'_1, \Delta'_2, \ldots, \Delta'_s$. Then there exist closed sets $\Delta^*_1, \Delta^*_2, \ldots, \Delta^*_s$, contained respectively in the regions of illumination $\Delta'_1, \Delta'_2, \ldots, \Delta'_s$ which, taken together, cover all the boundary of F.

Let us prove this assertion. Denote by Γ_i the boundary of the region of illumination Δ'_i (for example, in Figure 71, the boundary of the region of illumination consists of two points A and B; in the case of a three-dimensional convex body, the boundary of the region of illumination will be some line, and so on). Furthermore, define a function f_i on the boundary of the convex body F as follows. If A does not belong to the region of illumination Δ_i', set $f_i(A) = 0$. If F is a point of the region Δ_i', then let $f_i(A)$ be the shortest distance from A to the boundary Γ_i of the region Δ_i'.

It is clear that the function $f_i(A)$ is continuous and takes positive values on the points of the region Δ_i' (and only on these points). The sum:

$$\phi(A) = f_1(A) + f_2(A) + \ldots + f_s(A)$$

is a continuous function (given on the boundary of F), and taking only positive values (because each boundary point of F belongs to at least one region of illumination $\Delta_1', \Delta_2', \ldots, \Delta_s'$). Let σ be the smallest value of this function; thus, $\phi(A) \geqslant \sigma > 0$ for any boundary point A of the body F.

Now let us denote by Δ_i^* the set of all boundary points A of F for which the inequality $f_i(A) \geqslant \sigma/s$ holds. The set Δ_i^* is closed and is contained in the region of illumination Δ_i'. It remains to show that the sets $\Delta_1^*, \Delta_2^*, \ldots, \Delta_s^*$ cover all the boundary of F. Then $\phi(A) \geqslant \sigma$, or equivalently,

$$f_1(A) + f_2(A) + \ldots + f_s(A) \geqslant \sigma.$$

But then for at least one $i = 1, 2, \ldots, s$, the inequality $f_i(A) \geqslant \sigma/s$ must be satisfied and, consequently, the point A belongs to at least one of the sets $\Delta_1^*, \Delta_2^*, \ldots, \Delta_s^*$.

(12) (from page 59). In the n-dimensional case, there also exists a segment h_i such that the parallel translation of the set Δ_i^* in the direction ℓ_i' by a distance less than h_i translates Δ_i^* wholly *inside* the body F. In fact, for each point A of the set Δ_i^*, we shall denote by $g(A)$ the length of the chord being cut by F on the line parallel to the direction ℓ_i' and passing through A. The function $g(A)$, given on the set Δ_i^*, is continuous and takes only positive

values (because each point A of Δ_i^* belongs to the region of illumination Δ_i', and therefore the corresponding line passes through interior points of F). As the set Δ_i^* is closed and bounded, there is a point h_i in Δ_i^* at which the continuous function $g(A)$ takes its smallest value. Thus, $g(A) \geqslant h_i > 0$ for any point A of the set Δ_i^*. This means that for any line parallel to the direction ℓ_i' and passing through some point of Δ_i', the body F cuts a chord of length at least h_i. This implies our assertion.

(13) (from page 60). The final part of the proof of Theorem 7 goes through in the n-dimensional case in just the same way as in the main text, only that in this case we will not have "sectors" but "cones" $G_1, G_2, \ldots, G_s$ with apex O and curvilinear "bases" $\Delta_1^*, \Delta_2^*, \ldots, \Delta_s^*$.

(14) (from page 62). Let F be a convex n-dimensional body, having n corner points $A_1, A_2, \ldots, A_n$. Choose some directions $\ell_1, \ell_2, \ldots, \ell_n$ illuminating the points $A_1, A_2, \ldots, A_n$ respectively. If the direction ℓ_i "moves" slightly, the point A_i clearly remains a point of illumination for this direction, so we may suppose that the directions $\ell_1, \ell_2, \ldots, \ell_n$ are not parallel to a hyperplane. Draw from a point O some vectors $\overline{OB_1}$, $\overline{OB_2}$, $\ldots$, $\overline{OB_n}$, having directions $\ell_1, \ell_2, \ldots, \ell_n$, and let us construct from these points the vector

$$\overline{OB_{n+1}} = -\overline{OB_1} - \overline{OB_2} - \ldots - \overline{OB_n}.$$

We get $n+1$ points $B_1, B_2, \ldots, B_{n+1}$, and moreover, the point O lies in the interior of the simplex with vertices at these points. Therefore, the directions $\ell_1, \ell_2, \ldots, \ell_n$, defining the vectors $\overline{OB_1}$, $\overline{OB_2}$, $\ldots$, $\overline{OB_n}$, $\overline{OB_{n+1}}$, permit the illumination of all usual (that is, not corner) points on the boundary of F (see the proof of Theorem 9). The corner points $A_1, A_2, \ldots, A_n$ are also illuminated (by virtue of the choice of the directions $\ell_1, \ell_2, \ldots, \ell_n$). So if an n-dimensional body F has n (or fewer) corner points, then $c(F) = n + 1$.

(15) (from page 67). Let F be an n-dimensional unbounded almost conic convex body, not wholly containing any line, and K be the inscribed cone of this body. Let us suppose that the cone K has dimension q, and denote by L the hyperplane of dimension q, containing the cone K. Lastly, let us fix some point O in the hyperplane L, and draw through it a hyperplane P of dimension $n - q$, having orthogonal extensions to L. We shall denote by N the set of all points C of P for which it is possible to select a point A of the cone K and a point B of the body F such that $\overline{AB} = \overline{OC}$. The set N is an $(n-q)$-dimensional bounded convex body, which need not be closed. We shall denote the closure of N by M. This is an $(n-q)$-dimensional bounded convex body which, as shown by P.S. Soltan, satisfies $b(F) = b(M)$.

BIBLIOGRAPHY

1. Boltyansky V.G., The Problem of the illumination of the boundary of a convex body (in Russian). Izvestiya Moldavskogo filiala Akademii Nauk SSSR, *No.* 10 (76), 1960, 77-84.
2. Bonnesen T. and Fenchel W., Theorie der konvexen Körper, Springer, Berlin, 1934.
3. Borsuk K., Über die Zerlegung einer Euklidischen *n*-dimensionalen Vollkugel in *n* Mengen. Verh. Internat. Math. Kongr., Zürich, 2 (1932), 192.
4. Borsuk K., Drei Sätze über die *n*-dimensionale Sphäre. Fundamenta Math. *20* (1933), 177-190.
5. Danzer L., Überdeckungen mit kongruenten Kugeln und Durchschnittseigenschaften von Kugelfamilien in euklidischen Räumen hoher Dimension.
6. Danzer L. and Grünbaum B., Über zwei Probleme bezüglich konvexer Körper von P. Erdős und von V.L. Klee. Math. Zeitschrift, *79* (1962), 95-99.
7. Eggleston H.G., Covering a three-dimensional set with sets of smaller diameter. J. London Math. Soc. *30* (1955), 11-24.
8. Eggleston H.G., Convexity. Cambridge Univ. Press, Cambridge, 1958.
9. Eggleston H.G., Problems in Euclidean Space, Application of convexity, Pergamon Press, London, New York, 1957.
10. Erdős P., On sets of distances of *n* points. Amer. Math. Monthly *53* (1946), 248-250.
11. Erdős P., Some unsolved problems. Michigan Math. J. *4* (1957), 291-300.
12. Fejes Tóth L., Lagerungen in der Ebene, auf der Kugel und im Raum (2nd ed.), Springer, Berlin, 1972.
13. Gale D., On inscribing *n*-dimensional sets in an *n*-simplex. Proc. Amer. Math. Soc. *4* (1953), 222-225.
14. Gohberg I.Ts. and Markus A.S. A certain problem about the covering of convex sets with homothetic ones (in Russian). Izvestiya Moldavskogo Filiala Akademii Nauk SSSR, *No.* 10 (76), 1960, 87-90.
15. Grünbaum B., A simple proof of Borsuk's conjecture in three dimensions. Proc. Cambridge Philosophical Society *53* (1957), 776-778.
16. Grünbaum B., Borsuk's partition conjecture in Minkowski planes. Bull. Research Council Israel, *7F* (1957), 25-30.
17. Grünbaum B., On a conjecture of H. Hadwiger. Pacific J. Math. *11*, *No.* 1 (1961), 215-219.
18. Grünbaum B., Borsuk's model and related questions. Proc. Symposia Pure Math. *7* (1963), 271-284.
19. Hadwiger H., Überdeckung einer Menge durch Mengen kleineren Durchmessers. Comm. Math. Helv. *18*, (1945/46), 73-75; Mitteilung betreffend meine Note: Überdeckung einer

Menge durch Mengen kleineren Durchmessers. Comm. Math. Helv. *19* (1946/47), 72–73.

20. Hadwiger H., Über die Zerstückung eines Eikörpers. Math. Zeitschrift, *51* (1947), 161–165.
21. Hadwiger H., Ungelöste Probleme, N 20. Elem. der Math. 12 (1957), 121.
22. Hadwiger H., Über Treffanzahlen bei translationsgleichen Eikoerpen. Archiv Math. *8* (1957), 212–213.
23. Hadwiger H., Altes und Neues über konvexe Körper, Birkhäuser, Basel und Stuttgart, 1955.
24. Hadwiger H. and Debrunner H., Kombinatorische Geometrie in der Ebene, English Translation: Combinatorial Geometry in the Plane (with additions by V. Klee), Holt, New York, 1964.
25. Heppes A., Térbeli ponthalmazok felosztása kisebb atméröjü részhalmazok összegére. A Magyar Tudomanyos Akadémia Közleményei 7 (1957), 413–416.
26. Heppes A. and Révész P., Zum Borsukschen Zerteilungsproblem. Acta Math. Acad. Sci. Hung. *7* (1956), 159–162.
27. Klee V.L., Unsolved problems in intuitive geometry, Hectographical lectures, Seattle, 1960.
28. Lenz H., Zur Zerlegung von Punktmengen in solche kleineren Durchmessers. Archiv Math. *6*, *No.* 5 (1955), 413–416.
29. Levi F.W., Ein geometrisches Überdeckungsproblem. Archiv Math. *5* (1954), 476–478.
30. Levi F.W., Überdeckung eines Eibereiches durch Parallelverschiebungen seines offenes Kerns. Archiv Math. *6* (1955), 369–370.
31. Lyusternik L.A., Convex sets and polyhedra (in Russian), Gostekhizdat, Moscow, 1956, (English translation, Dover, New York 1963).
32. Lyusternik L.A. and Schnirel'man L.G., Topological Methods in Variational Problems (in Russian), Trudy of the Scientific Investigations Institute of Mathematics and Mechanics, Moscow, 1930.
33. Pál J., Über ein elementares Variationsproblem, Danske Videnskab. Selskab., Math.-Fys. Meddel. *3*, *No.* 2 (1920).
34. Soltan P.S., The illumination of the boundary of a convex body from within (in Russian). Matem. Sbornik (novaya seriya), *57* (99) (1962), 443–448.
35. Soltan P.S., Towards the problem of covering and illumination of convex sets (in Russian). Izvestiya Akademii Nauk Moldavski SSR, *No.* 1 (1963), 49–57.
36. Visityei V.N., Problems of covering and illumination for unbounded convex sets (in Russian). Izvestiya Akademii Nauk Moldavski SSR, *No.* 10 (88), 1962, 3.
37. Yaglom I.M. and Boltyansky V.G., Convex sets (in Russian), Gostekhizdat, M., 1951.